Hairy-foot

long-tongue

Published by

Whittles Publishing Ltd,
Dunbeath,
Caithness, KW6 6EG,
Scotland, UK

www.whittlespublishing.com

ISBN 978-184995-564-5

Printed and bound in the UK
by Halstan Printing Group, Amersham.

Hairy-foot, long-tongue

Solitary bees, biodiversity & evolution in your backyard

David J Perkins

Whittles Publishing

Hairy-foot

To Mary Jo

Anthophora plumipes *(the "feathery-footed flower gatherer") : the Hairy-footed flower bee : the hind leg of the female; the head of a male from below.*

long-tongue

Solitary bees, biodiversity & evolution in your backyard

This is the hairy foot...

of the Hairy-footed flower bee...

and this is its very, very long tongue

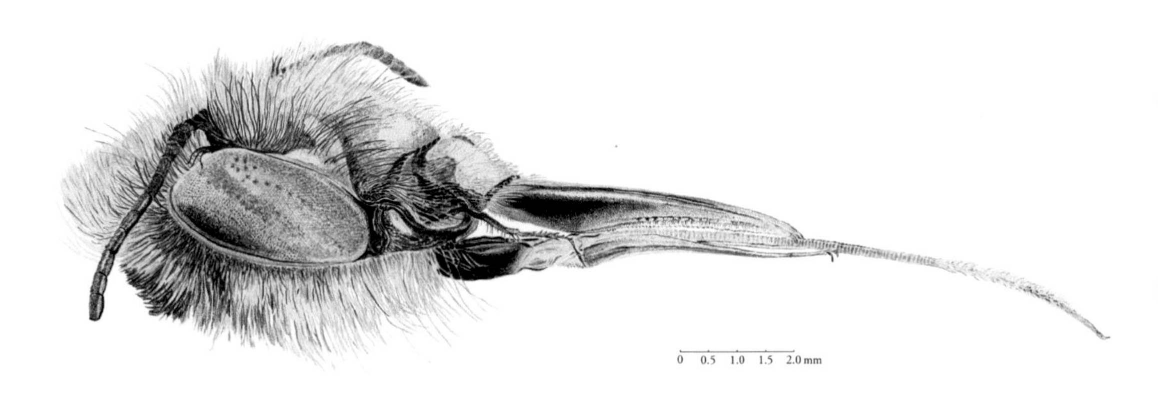

The mid-leg and head of a male Hairy-footed flower bee - Anthophora plumipes

A female Anthophora plumipes *approaching the flowers of a cowslip*

Why is it hairy and why is it long?

CONTENTS

Publisher's note: this highly illustrated book utilises double page spreads to present images and text. To enable the reader to get the most from this style, we have strayed from convention and page 1 appears as the first left-hand page.

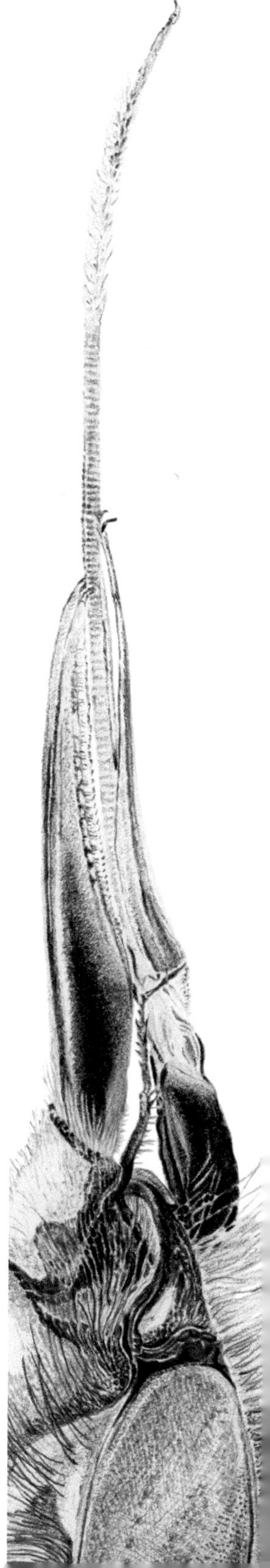

The hairy foot of the hairy-footed flower bee – and its very long tongue – are the starting points for this exploration of the biology and adaptations of 'solitary' bees (as opposed to 'social' bees that nest in large colonies). The book began with the drawing of an *Anthophora plumipes* female visiting a cowslip. Dead females, found on people's hearths, had been brought to me to solve the mystery of how they got there and what they were. The bees must have emerged from nests in chimney stack mortar and fallen down to the fireplace, weak and disorientated. Another case involved 'strange bees crawling around under the bed' (there was a fireplace in the bedroom). I had already become very fond of *Anthophora* in the wildlife garden I managed and was then able to study the bees under a microscope. They are quite large so dead individuals can often be found in gardens and I was lucky to find a dead male with his mouthparts extended. All the drawings in the book have been made from similar samples, or from digital photography and video (I rarely try to capture bees, only seek, find, observe, record). This is not a comprehensive book of bee biology or identification: there are many of these already, especially on bumble and honeybees. It aims primarily to deepen your enjoyment of the lives of the bees around you by illustrating the wonders of bee diversity, complexity and evolution. If you wish to explore further you will find detailed notes on sources, references and further reading arranged by page number at the end of the book.

Why are bees hairy and why do some of them have such long tongues? With these questions in mind we will look at some of the bees that live in our gardens. We will discover aspects of their amazing biology, such as how they sense the world or how they achieve their impressive aerobatics (right). In addition, we will interrogate time – how did they get here, just outside the door, with the highly evolved adaptations that make them the varied and successful insects we see today?

Our main character is the hairy-footed flower bee *Anthophora plumipes*, a round, hairy and very cute bee that flies in the early spring. The female hairy-footed flower bee (above) is an elegant bee coloured black and dark brown that may be easily mistaken for a worker bumblebee. The male (left) is a handsome tawny-and-ginger bee with pale yellow patches on his face. They fly with a fast, swerving action, with pauses to hover – either in front of flowers or, in the case of the male, when looking for females. The male has particularly hairy feet and a distinct fan on his middle legs.

The hairy-footed flower bee is just one of many bee species you might find in your garden. Unlike bumblebees and honeybees, most of these bees have a solitary lifestyle. The females make nests alone and the young in the nests feed themselves on food left by the mother. The females die after a relatively short season of nest-building and the young emerge from the nest hole long after the females have disappeared – in the following spring or, for a few species, later in the same summer. Males die after a short, intense season of finding females and competing for mating, which usually takes place shortly after new females have emerged.

The hairy-footed flower bee generally emerges in March when the first chiffchaffs are singing in the blackthorn crowns, though in London I have seen it feeding on winter-flowering honeysuckle in February. Depending on the weather, the females will then be busy until, perhaps, the beginning of May, when the first swifts are arriving high above the city. Then they disappear and there is a long wait to see their cheery, swooping flights again the following spring.

As the year progresses different species of solitary bee emerge from their nests and appear in our gardens. With good gardening we can expect a rich succession of species through the spring and into autumn. We meet some of these other bees. There is guidance for your gardening practice later in the book.

It is true that you will find a greater diversity of bees if you live in the south of England, but with climate change and milder winters many of the species doing well are expanding their ranges northwards. The hairy-footed flower bee was reported in Edinburgh in 2015 and in Kilmartin, Argyll in 2020. It reached Ireland in March 2022. Expect interesting times ahead; there will be losses and sadness but there may also be new bees in your gardens in the coming years.

These same strange winters have led in recent years to an increasing number of bumblebee nests being started by new queens in the autumn. The workers from these nests can be found gathering pollen and nectar from winter-flowering shrubs such as *Mahonia* and winter-flowering honeysuckle *Lonicera fragrantissima*.

So, certainly in some parts of the country, we could say there is:

'A bee for all seasons'

Bees, wasps and ants are related and are placed in the same order of insects, the Hymenoptera (bees, wasps, ants and sawflies – insects with membranous wings).

Bees, Wasps & Ants

Clockwise from top left: the honeybee Apis mellifera*; the ornate-tailed digger wasp* Cerceris rybyensis*; the parasitic greater pennant wasp* Gasteruption jaculator*; the yellow meadow ant* Lasius flavus.

The number of Hymenoptera species is huge. There are at least 150,000 known worldwide and more are being found. There may be more than one million in total. There are around 20,000 species of bee alone. Relationships between the many species are complex and the understanding of them is a developing science (the discipline of phylogenetics). A broad classification of groups in the Hymenoptera is shown opposite.

The first major division in the order is between the Symphyta and the Apocrita. Symphyta includes sawflies, some of which are well-known garden pests. They have a simpler body form with no 'wasp waist' between thorax and abdomen. The Apocrita, the ants, bees and wasps, are then split into Aculeata and Parasitica (though this latter name is now replaced – see Notes, p.152). The aculeate Apocrita all have stings whereas in the Parasitica the sting has not evolved. The latter are often called 'parasitic wasps' (though note that parasitism is also frequent in the Aculeata, pp.103-110). Within the Aculeata are the 'super-families' Apoidea (including the bees, Anthophila, and the 'apoid wasps'), Vespoidea (wasps and ants) and Chrysidoidea (cuckoo wasps). For more discussion of the evolution of the Hymenoptera see p.47–68 and the Classification section p.141.

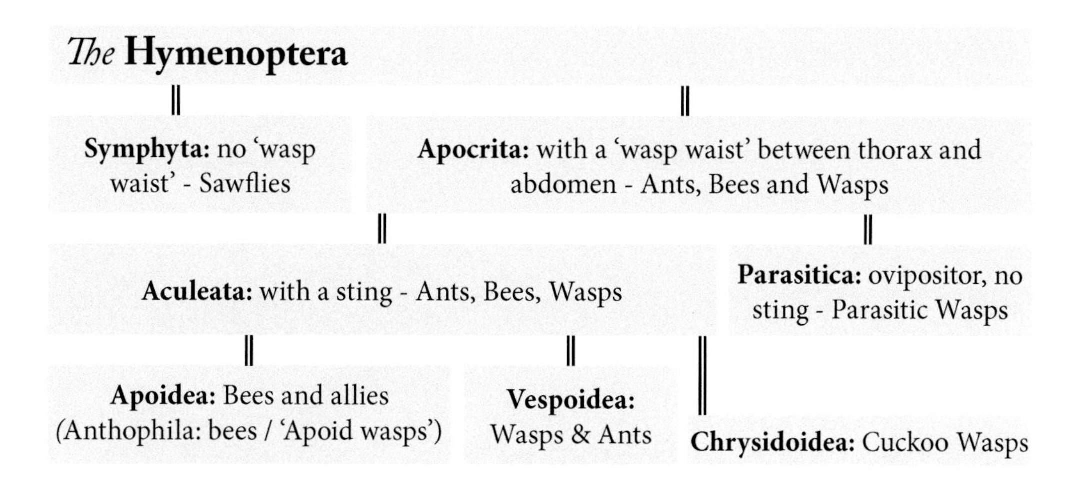

If Earth is to be thought of as 'a planet of...', it should perhaps be a 'planet of insects'. Of the insect groups the Hymenoptera contribute a very significant proportion of known, named species (13%). The Lepidoptera (butterflies and moths, 16%) and the Diptera (flies, 12%) are of similar importance. Only the Coleoptera (beetles) play a more significant role. As more research is done on the diversity of the innumerable tiny parasitic wasps, however, the dominance of Coleoptera is being challenged.

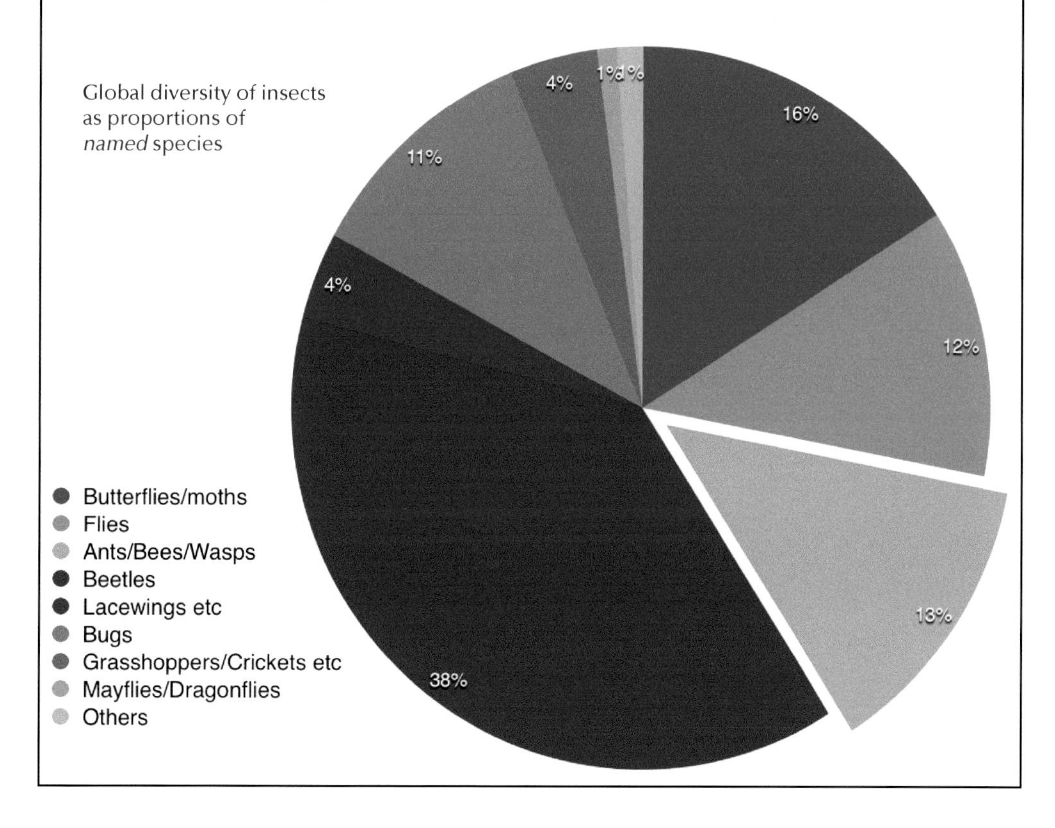

Numbers of species in the British Isles

Genus	No.
Andrena	68
*Anthidium**	1
Anthophora	5
Apis	1
Bombus	27
Ceratina	1
Chelostoma	2
Coelioxys	7
Colletes	9
Dasypoda	1
Dufourea	2
Epeolus	2
Eucera	2
Halictus	8
Lasioglossum	33
Heriades	2
Hoplitis	2
Hylaeus	12
Macropis	1
Megachile	7
Melecta	2
Melitta	4
Nomada	36
Osmia	13
Panurgus	2
Rophites	1
Sphecodes	18
Stelis	5
Xylocopa	1
TOTAL	275

* Now 2 species - see p, 119

The violet carpenter bee Xylocopa violacea *(here a female), is the largest bee in Europe and has a forewing length of 22mm. Looking on is one of the smallest bees, a male least furrow bee* Lasioglossum minutissimum*, which has a forewing length of only 3mm.*

There are around 275 bee species in the UK, though some may be extinct. There are 1,965 species in Europe of which 9% are threatened with extinction (IUCN, 2014). New species are regularly discovered: two new breeding species, *Hoplitis adunca* and *Nomada alboguttata*, were added to the British list in 2016; three more, *Nomada facilis, Osmia cornuta* and *Heriades rubicola*, were added in 2017; *Stelis odontopyga* and *Nomada bifasciata* were confirmed in 2018; *Sphecodes albilabris* was confirmed in 2020. In 2018 the mining bee *Andrena proxima* was split, adding *Andrena ampla* to the list. *Nomada panzeri* and *Epeolus cruciger* are also likely to be split. In 2021 a new species of *Anthidium* was discovered in the UK (p.119). Seven *Anthophora* species new to science were described in the US in 2016 and one in Kurdistan in 2017. Most species of bee do not live in large, social nests: they are 'solitary'. Unlike the highly social bees with young females becoming workers to raise later generations, solitary bee nests are stocked with enough food and then sealed for safety. The larvae grow in the nest on their own.

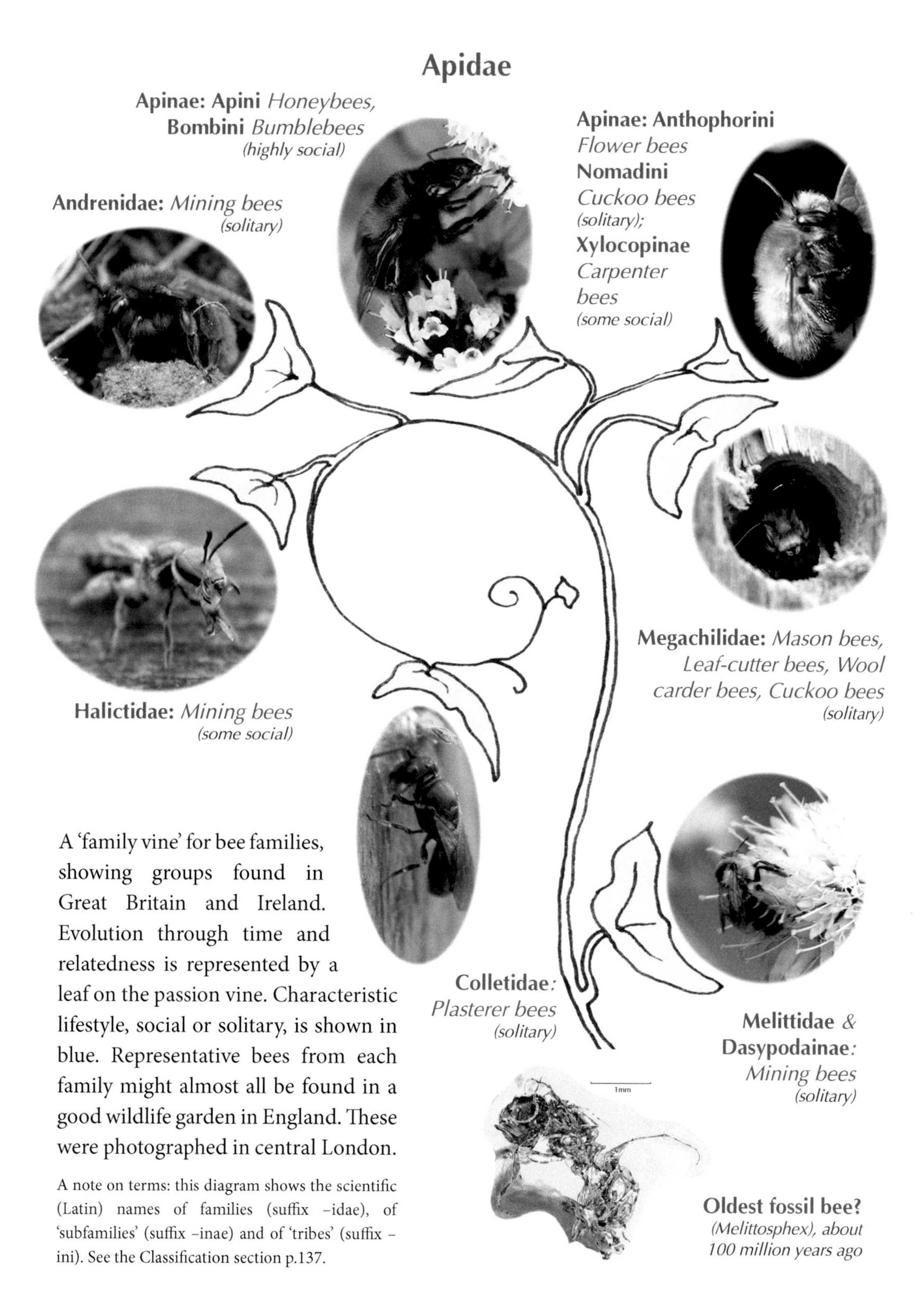

A 'family vine' for bee families, showing groups found in Great Britain and Ireland. Evolution through time and relatedness is represented by a leaf on the passion vine. Characteristic lifestyle, social or solitary, is shown in blue. Representative bees from each family might almost all be found in a good wildlife garden in England. These were photographed in central London.

A note on terms: this diagram shows the scientific (Latin) names of families (suffix –idae), of 'subfamilies' (suffix –inae) and of 'tribes' (suffix –ini). See the Classification section p.137.

The adult bee's body has three primary components, the head, thorax and abdomen, which are supported by a segmented exoskeleton.

The **head** holds the brain, sense organs (two large, compound eyes; three small eyes – ocelli – in the top-centre of the head; two antennae), the mouthparts and important glands. The complex compound eyes give the bee information about shape, movement and colour. The antennae are mobile organs highly sensitive to scents and touch. The mouthparts of bees are complex and enable them to cut and mould materials using the mandibles and to gather liquids such as nectar and water using the tongue (glossa). The delicate glossa is surrounded above by the galeae (one galea on each side) and by two labial palps beneath. See p.39 for more detail.

The **thorax** supports the wings and legs and contains the muscles. It is the 'power box' of the bee.

The **abdomen** contains the digestive system, reproductive organs and important glands, including the venom gland and sting in the female.

The three parts of the body are linked by the nervous system, the oesophagus and digestive system, the tracheal (breathing) system and the heart and circulatory system.

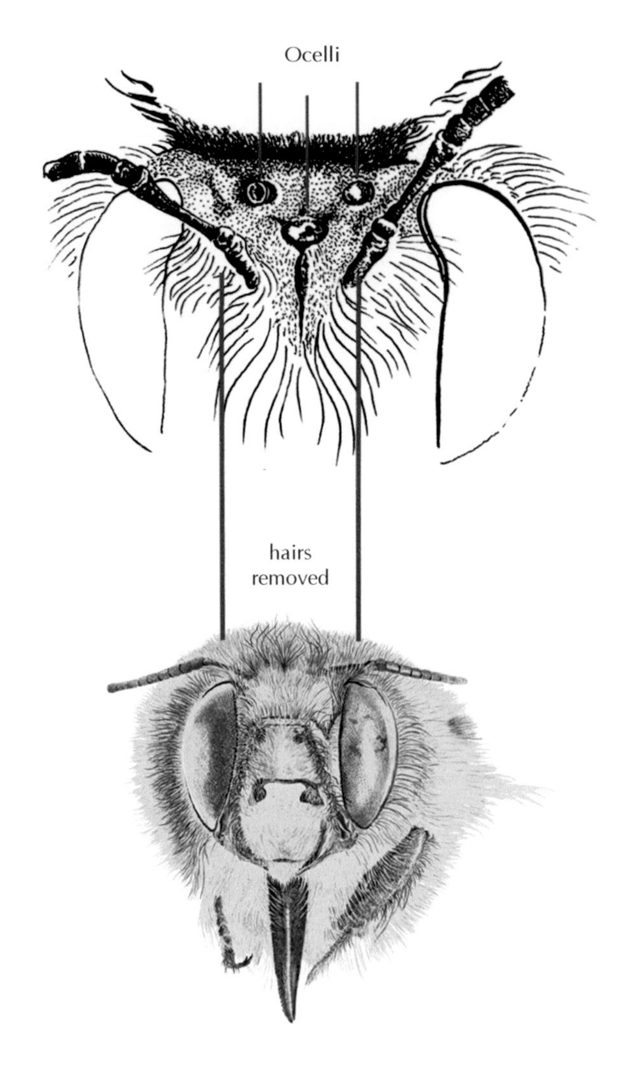

The ocelli (singular ocellus) are 'simpler' eyes than compound eyes and give the bee information about light intensity and therefore time of day and weather conditions. They are also essential for balancing during flight (the sky is the brightest part of the environment). The ocelli are just as important to a bee's survival as the more prominent compound eyes. This is particularly the case for the nocturnal neotropical bee *Megalopta genalis*, which has enlarged ocelli that provide more sensitive dim-light vision in its forest habitat.

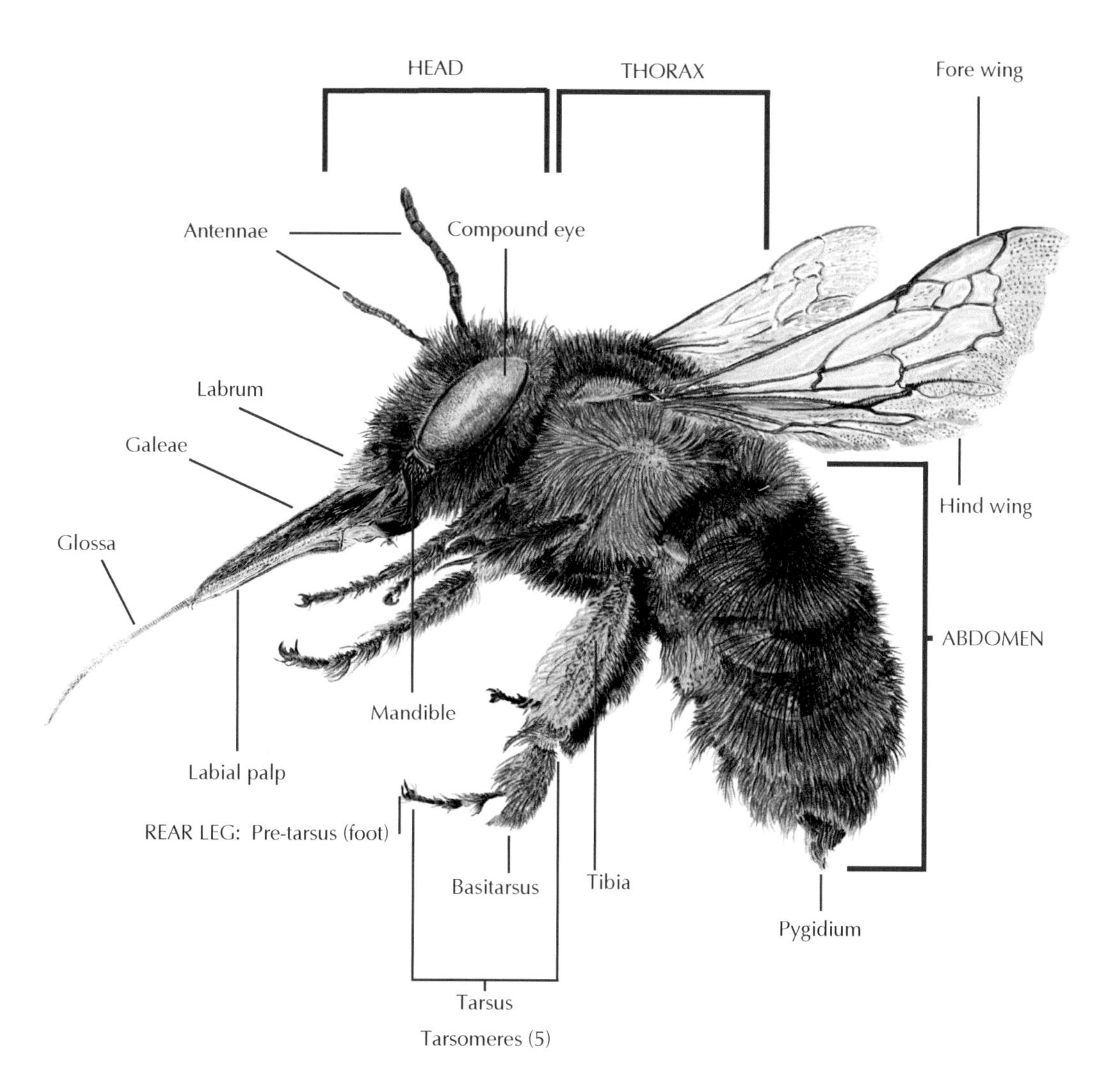

Some external anatomical terms for bees shown by Anthophora plumipes.
More details can be found in later sections.

The larval body plan is much simpler: it is an efficient feeding tube, though it also contains the complex potential for metamorphosis into an adult (pp.51-54).

Evolution over 100 million years has produced the special characters of contemporary bees. Bees have numerous adaptations to their hairs, tongue length, and glandular forms and functions.

As insects, bees generally need to sunbathe to raise their temperatures (they are **ectothermic** or cold-blooded; their body temperatures are linked to the environment) but many bees can raise their temperatures above that of the surrounding air. This makes them, at times, **endothermic**, but because this is temporary and behavioural, they are **heterothermic**. Flower bees and bumblebees are particularly good at raising their own temperatures, and so can fly on cool spring days.

To understand this think about the body of the bee. It has three sections separated by narrow constrictions. Between the thorax (Th, below left) and the abdomen (A) there is the petiole (P).

The blood of insects, called **hemolymph** (light yellow, below), 'bathes' the organs and muscles and is not restricted to veins or arteries. It is the main carbohydrate store in insect muscles. There is a heart – a long, narrow organ beneath the upper surface of the abdomen. Hemolymph enters the heart in the abdomen through valves and is pumped forwards through the thorax into the head, where it bathes the head glands and brain. It can enter the antennae and mouthparts and is used partly to extend and move them. The pulsing of two diaphragms in the abdomen produces a circulation that draws the hemolymph back through the thorax into the abdomen. This circulation delivers nutrition to muscles and organs and is crucial in temperature control.

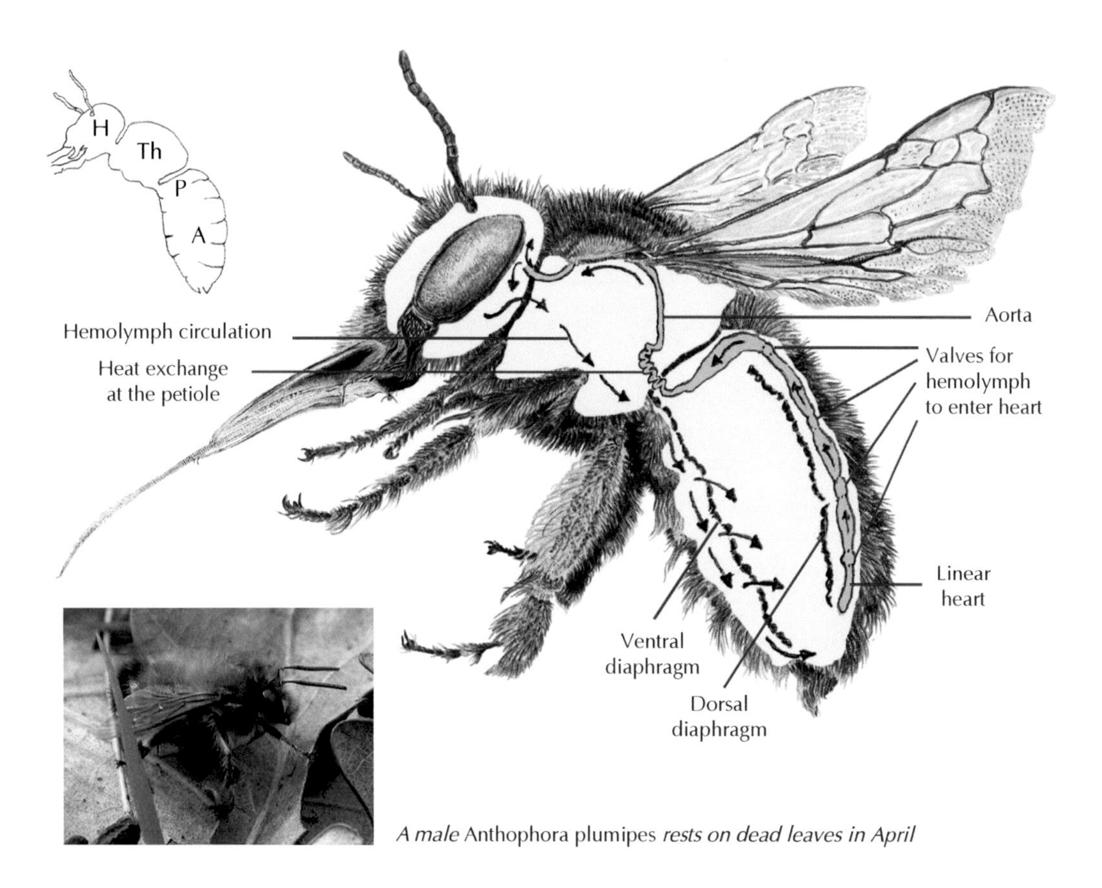

A male Anthophora plumipes *rests on dead leaves in April*

Anthophora can get on, without waiting for 'hot air'

Heat is generated by muscle activity in the thorax, which contains all the large muscles. Flight muscles are particularly prone to heating up and on a hot day it is crucial for bees to cool down. This is one job of the hemolymph, which carries heat from the thorax back into the abdomen, where it can cool. It can then be returned via the heart to produce an effective cooling system. Heat can also be lost by bringing fluids from the digestive system to the mouth to evaporate (like a dog panting). If, however, the bee is cold and needs to warm up for flying, this cooling system must be switched off. Bees can tremble their wing muscles rapidly to generate heat (just like shivering, though it is not visible). To stop heat escaping and raise the bee's thorax to flying temperature quickly, heat is transferred at the petiole to the cooler hemolymph: it is an effective heat-exchange system to counter the cooling system. Honeybees are better than flower bees at keeping their bodies cool when flying on hot days, but flower bee heat exchange is better. Flower bees can tolerate lower air temperatures in flight and can fly when their thorax reaches 25°C, whereas honeybees must wait until their thorax reaches 28°C or higher (27°C in emergency take-off). The rate at which larger bees warm up can be 10–15°C per minute, which is faster than the rate of many mammals.

The hairs of all bees are **plumose** (branched, see below). In hairy bees such as *Anthophora* or bumblebees a thick coat of plumose hairs creates an efficient insulation layer. Flower bees and bumblebees can often be seen flying early on cool spring mornings, even in frost, when other insects remain comatose – they need not wait for the air to heat up. See p.23 for a second, crucial function of the plumose hairs.

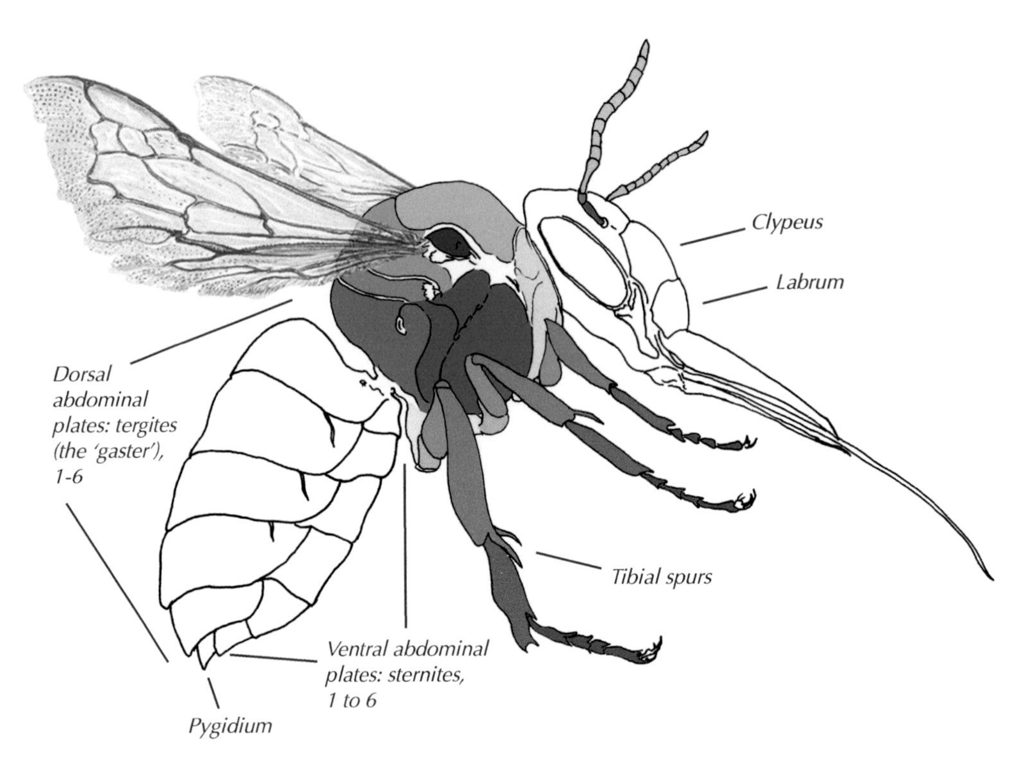

anti-clockwise around the thorax

THORAX
Pronotum
Mesoscutum
Mesoscutellum
Metanotum
Propodeum
Episternum/ Mesepisternum
Tegula

ANTENNAE
Scape
Pedicel
Flagellum

LEGS
Coxa
Trochanter
Femur
Tibia
Basitarsus
Tarsus

Beneath the hairs of *Anthophora* is an **exoskeleton** that comprises a cuticle (composed of chitin) that is segmented and jointed. Chitin can be hardened to form the protective plates of the exoskeleton (tergites, sternites and other plates of the thorax) or the very tough mandibles, or it can be flexible and membranous to form links between plates. Hardened chitin is waterproof and often waxy on the outside, preventing loss of body fluids in dry conditions. The complexity of the exoskeleton is simplified above and is based upon *Anthophora*. The features shown are useful when thinking about the evolution and behaviour of bees. The thorax, for example, which is separated from the abdomen by the petiole, contains a segment – the **propodeum** – that in the ancestral insect was the first segment of the abdomen. The thoracic and abdominal plates are penetrated by openings – **spiracles** – that allow air to enter the respiratory system (p.15). In female *Anthophora* and other mining bees the sixth abdominal tergite is the **pygidium** (see p.46), which is used in nest-making to compact soil and spread waterproofing secretions.

Glandular secretions are crucial to bees. The generalised diagram above shows the glands and digestive system of bees. The development and function of the many glands in a bee's body vary across species according to their adaptations. In mining bees such as *Anthophora*, the Dufour's gland is large and its secretions are used to line underground nest cells, waterproofing the cell and discouraging bacterial and fungal growth. It is also nutrition for the larvae, which consume the lining (see p.53). In other species, Dufour's gland secretions are used for nest-site recognition. In social bees such as bumblebees or *Lasioglossum* species it is also used for nest-mate recognition and fertility signalling. In the highly social honeybee, the Dufour's gland is smaller and its secretion is one part of the queen's complex pheromone signal to the thousands of workers in the hive. Other glands such as the salivary and hypopharyngeal glands also have a range of functions across different species. Wax glands are only present in highly social honeybees, bumblebees and tropical stingless bees.

At times it can feel like we are looking at aliens: the evolution of life has given us endless solutions for being. In the insect nervous system there is an alternative to our vertebrate 'big brain'. The brain in the bee, though small, does achieve an immense amount of complex behaviour. But there is also a 'distributed' aspect to the bee's nervous system. There are prominent ganglia (a **ganglion** is a cluster of nerve cells) along the ventral surface (underside) of the bee's body (see below, 1g–7g). These control bodily functions independently of the brain; the thoracic ganglia and associated nerves operate the wings and legs. A decapitated insect thorax and abdomen can continue to operate in a simpler way for some time.

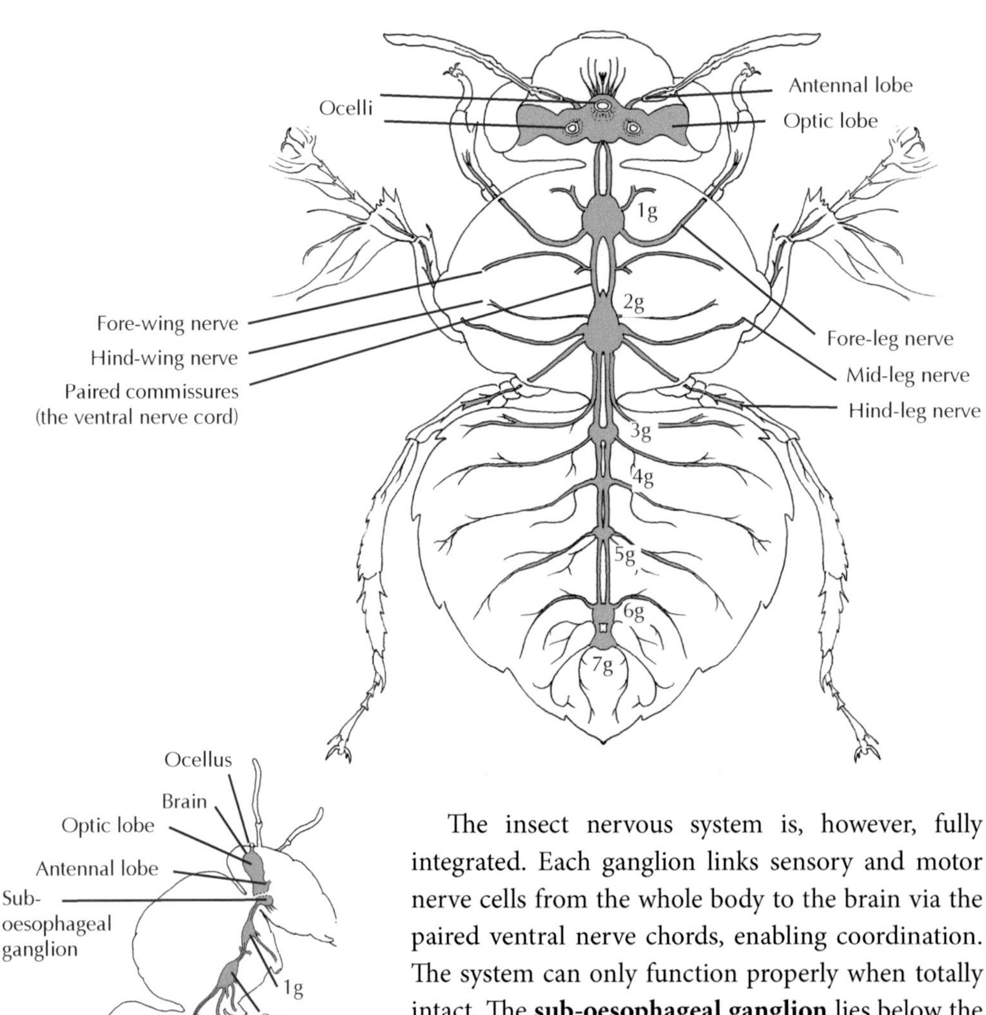

The insect nervous system is, however, fully integrated. Each ganglion links sensory and motor nerve cells from the whole body to the brain via the paired ventral nerve chords, enabling coordination. The system can only function properly when totally intact. The **sub-oesophageal ganglion** lies below the brain and the oesophagus, which has to pass through the brain to the mouth. This arrangement of brain and ganglia originates during larval development and although altered they are not re-absorbed during metamorphosis.

Apian intellect - sophisticated, even if not abstract!

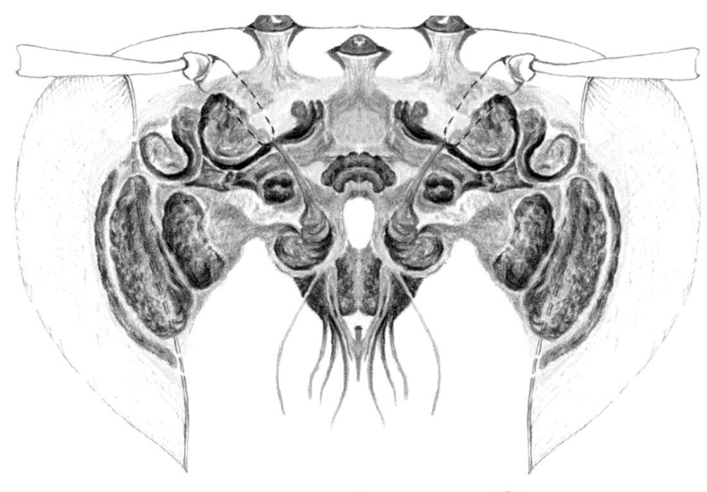

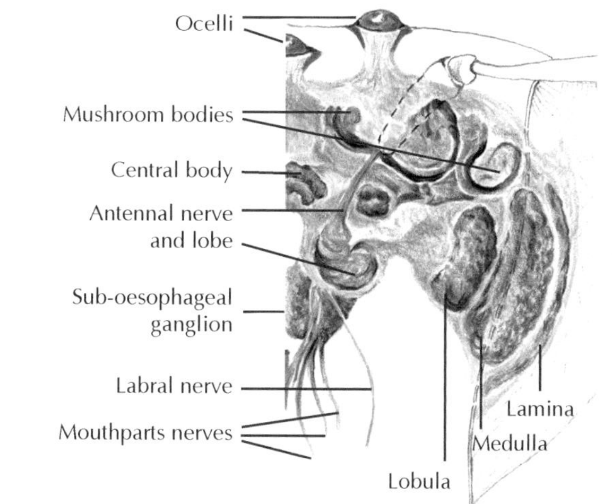

The bee's brain, an elegant structure, provides enough memory and processing power (just under one million neurons in the honeybee; humans have 100 billion) to deal with complex sensory output from the optic and antennal lobes – the compound eyes, ocelli, antennae and mouthparts. Despite its tiny size it can also integrate information from the thoracic and abdominal ganglia that enable the bee's complex behaviour, such as the learning and memorising of landscape features that facilitate successful flights from nest to feeding areas, the use of chemical signals produced during feeding and mating strategies (e.g. pp.34, 36) or coping with weather and short-term environmental change during nest-building, stocking and defence (e.g. pp.56, 82).

The large **mushroom bodies** of the brain are where important spatial and temporal learning and memory takes place. They also enable the evolution of eusocial behaviour (advanced social organisation; pp.57-64). The **central body** is responsible for 'dead-reckoning' abilities, by which the details of an outbound journey (including the use of polarised light as a compass and of optic flow as a measurement of flight speed and time) are integrated. This means the shortest route can be taken back to the nest, however tortuous the outbound flight. The **optic lobes**, each consisting of the lamina, medulla and lobula, feed vital information from the compound eyes to the brain; the **antennal nerves** and lobes do the same for the complex signals coming from the antennae. The **sub-oesophageal lobe** controls the mandibles, the hypopharynx, the maxillae (including maxillary palps and galeae) and the labium (including the mentum, labial palps and glossa; p.39). Large mushroom bodies probably developed early in the evolution of the Hymenoptera at the onset of parasitoid lifecycles. Bees have small brains but over 140 million years have evolved the complexity needed for a complex life.

Insects have a much less centralised respiratory system compared to mammals. Rather than a couple of orifices, a tube and a pair of lungs transferring oxygen to a liquid transportation mechanism (blood), insects use a number of openings and a complex network of air passages to transfer oxygen directly to muscles and organs. The oxygen moves through increasingly narrower passages – the **tracheae** – to reach the tissues. The blood of insects (hemolymph, p.9) is not used for transporting oxygen. Instead, air enters the animal through the **spiracles**. In bees there are ten spiracles on each side – two on the thorax, one on the propodeum and the remainder along the abdomen (below). There are no respiratory openings on the head.

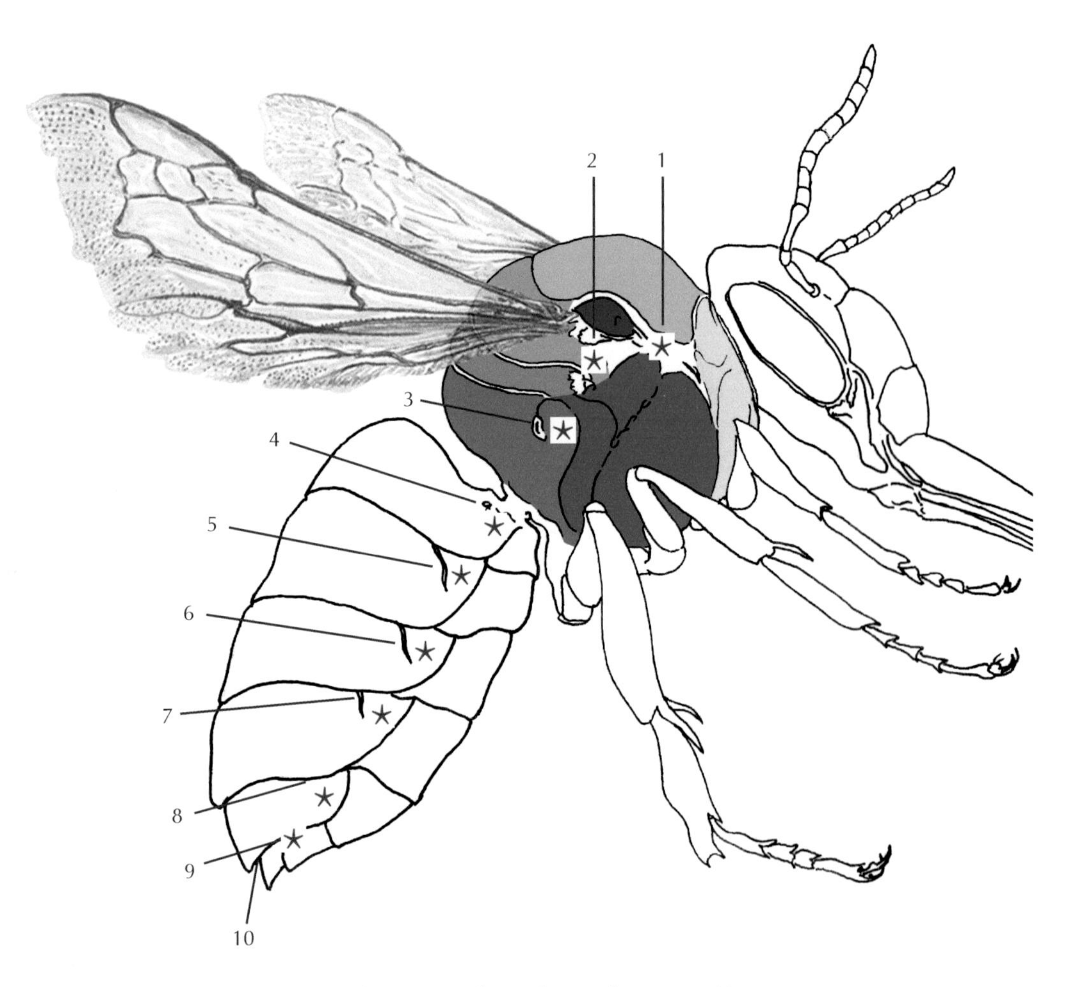

The location of spiracles in bees. The tenth spiracle is not visible.

Below, the insect respiratory system with open air to the left: s denotes the spiracle; fh the filter hairs; at the atrium; v the valve closure; tt the tracheal trunk; as the air sac; t2 and t3 the tracheal branches; tr the tracheoles.

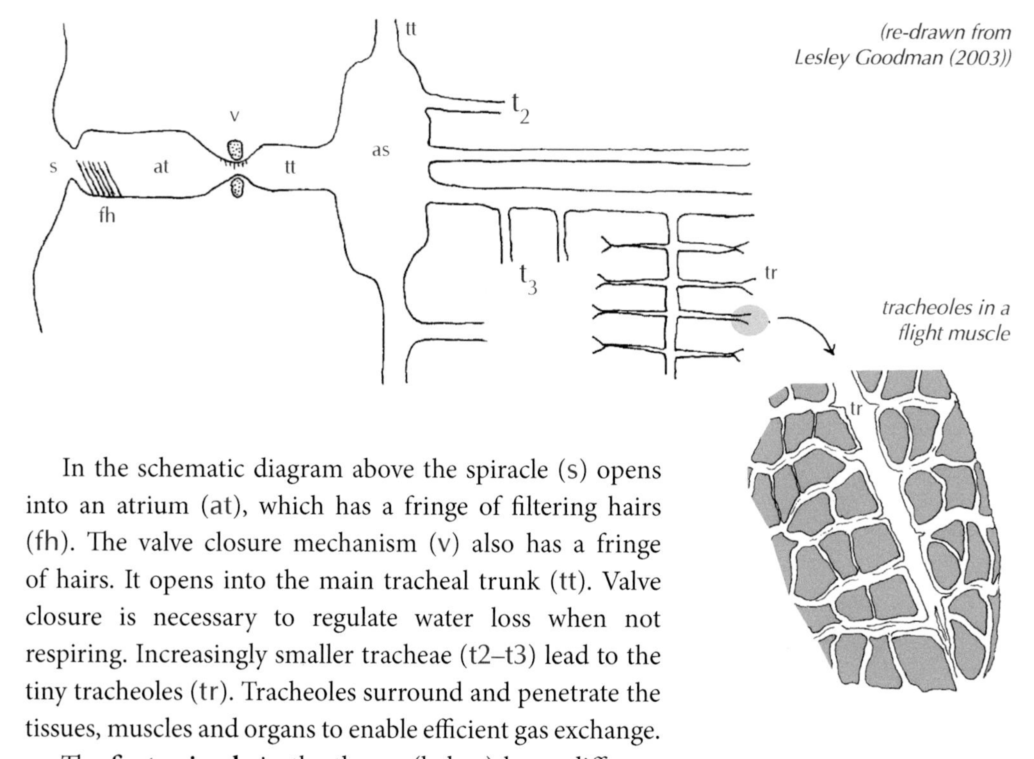

(re-drawn from Lesley Goodman (2003))

In the schematic diagram above the spiracle (s) opens into an atrium (at), which has a fringe of filtering hairs (fh). The valve closure mechanism (v) also has a fringe of hairs. It opens into the main tracheal trunk (tt). Valve closure is necessary to regulate water loss when not respiring. Increasingly smaller tracheae (t2–t3) lead to the tiny tracheoles (tr). Tracheoles surround and penetrate the tissues, muscles and organs to enable efficient gas exchange.

The **first spiracle** in the thorax (below) has a different closure mechanism to the spiracles in the abdomen. It has an external lobe and operculum with muscle attachments and no atrium, and a fringe of filtering hairs on the spiracular lobe. In highly active insects such as bees, with advanced flying skills, there are also large **air sacs** that enhance air intake and supply (next page).

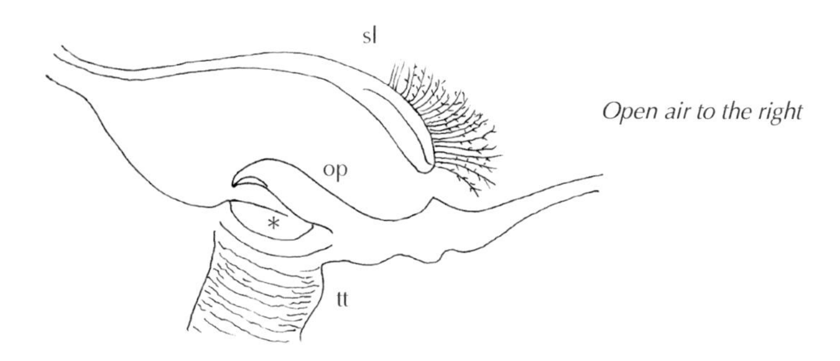

*Above: simplified diagram of the first thoracic spiracle: sl is the spiracular lobe; op the operculum; * the spiracular opening and tt the tracheal trunk.*

The **second spiracle** on the thorax is small and has no closure device.

The **third spiracle**, on the propodeum, is large, oval in shape and very visible. Its closure mechanism is different again (below). The cuticular rim of the opening and the atrium (with no filtering hairs) support a long oval valve with lobes at each end for the attachment of a closure muscle. As this contracts, the valve becomes more convex, its margin moving across the tracheal aperture to slot into a groove in the cuticular rim, thus closing the spiracle. A separate opener muscle attached to the propodeum wall acts in opposition, so that when the closure muscle relaxes the valve can be moved back from the groove, opening the aperture once more. The **abdominal spiracles** of bees have a valve mechanism similar to that shown in the general schematic diagram on the previous page.

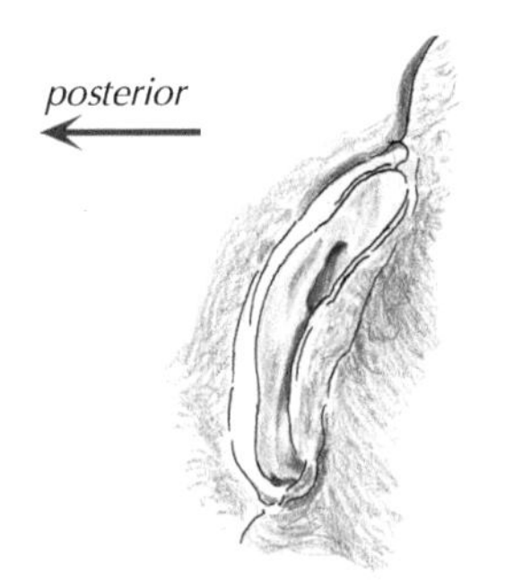

The third spiracle in Anthophora plumipes *(hairs removed).*

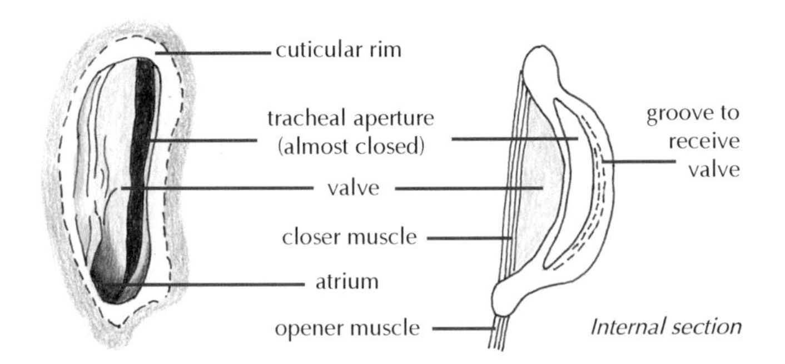

Honeybee third spiracle (re-drawn from Goodman, 2003)

Insect flight muscles are some of the most active of all animal tissues and so require very efficient supplies of oxygen. This is particularly true of bees – think of the hovering and manoeuvring skills of *Anthophora*. Many of the tracheae in bees are expanded into air sacs (a1–10, right), creating a considerable volume of air available for passage into the organs and muscles. There are large sacs over the brain (a1) the optic lobes (a2) and lower brain and mouthparts (a3). These are fed from large tracheal trunks leading from spiracle pair 1 (s1) (there are no spiracles in the head). These spiracles, along with s2 and s3 also feed a series of other important air sacs in the thorax (a4–a8). From the air sacs very large numbers of subdividing tracheae and even finer tracheoles penetrate the brain tissues to ensure a rich supply of oxygen to the nervous system. The thoracic air sacs (a4–a8) surround all leg and, especially, flight muscles. Two large tracheal trunks pass through the petiole to two enormous air sacs in the abdomen (a9, a10), themselves fed by the abdominal spiracles and linked by transverse commissures – shown in pink. The complex network of interconnecting tracheae and the sophisticated use of the closure valves in the spiracles enable the bee to take in sufficient oxygen for and expel CO2 from its highly active muscles and nervous system. Air sacs coloured pink lie ventrally unlike those in yellow.

Movement of air through the spiracles and air sacs is driven by rapid changes in the abdomen and thorax, especially during flight. The amount of overlap of the abdominal

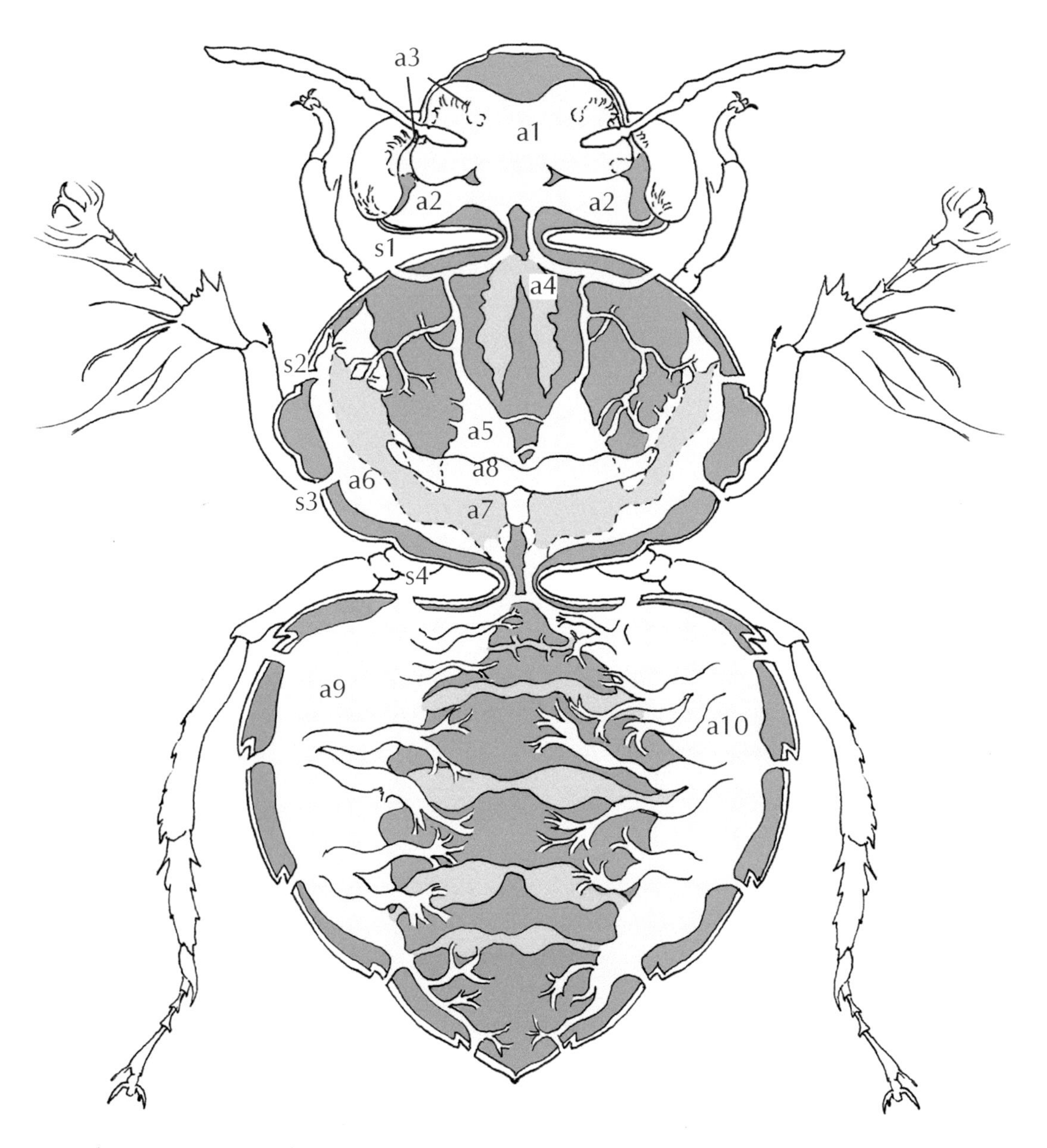

A bee's respiratory air sacks and some tracheae, which in detail vary from bee to bee.

tergites and sternites can be increased or decreased by muscles attached to the tergal and sternal plates. This lengthens or shortens the abdomen and, with muscles attached vertically from tergal to sternal plates, helps to flatten the abdomen. Opposing muscles expand it again vertically, achieving a pulsating volume change for the whole abdomen. This compresses then relaxes the air sacs, drawing in air and ejecting CO2.

These few, simplified diagrams introduce the wonders and complexities of insect flight. Its evolution has consumed many books, and deserves another. The thorax is the power unit for the bee. It contains huge muscles and associated air sacs, ganglia and nerves. The muscles of mammals are anchored on a more or less rigid endoskeleton but bee flight muscles are attached to the inside of the exoskeleton. This, being composed of plates joined by flexible cuticular membranes, can distort during muscle action. The rapid movements of the wings are largely achieved using **indirect** muscle action via the **dorso-ventral** muscles shown in pink opposite and the **longitudinal** muscles marked orange. On contraction, these muscles change the shape of the thorax rather than moving the wings, so raising or lowering the wings. In diagram C, the contracted dorso-ventral muscles and relaxed longitudinal muscles have pulled the roof of the thorax down, elevating the wings. In diagram D, the longitudinal muscles are contracted and the dorso-ventral muscles are relaxed, raising the roof of the thorax and depressing the wings. **Direct** muscles (pale yellow) are attached to projections – **sclerites** – at the bases of the wings and provide detailed rotational and steering control for the complex motions of the wings. The indirect muscles are stretch-activated – once in motion the opposing pair continue to contract and relax until stopped by the nervous system. This achieves much faster wing beats than a system relying entirely on nervous system control, as in birds and mammals. The flying skills of bees such as *Anthophora* also require a finely tuned sensory and nervous system of information gathering, processing and reactions to control the muscles and wings. Vision is clearly a key sense, but nerve cells associated with hairs on adjacent parts of the head, thorax, antennae and wings also play a major role (see pp.37–38).

As well as the powerful, rapid wing beats required for sustained flight using indirect musculature, the manoeuvrability displayed by bees also needs fine directional control provided by the direct muscles. This will affect pitch ('nose-up' or 'nose-down' rotation), roll (rotation about longitudinal axis, tilt to right or left) and yaw (rotation about vertical axis, nose-left or nose-right) during flight. This control enables the female hairy-footed flower bee to aim her glossa straight for a narrow opening for nectar before any actual landing or pause in flight. Males that defend territories, approaching, inspecting and driving off other males, require excellent hovering and avoidance skills. The energy expenditure involved in all of this (for a flying bee approximately 500 watts per kilogram, compared with the maximum output of an Olympic rowing crew of about 20 watts per kilogram) determines flight behaviour - foraging females and males defending resource-rich territories don't waste either time or energy.

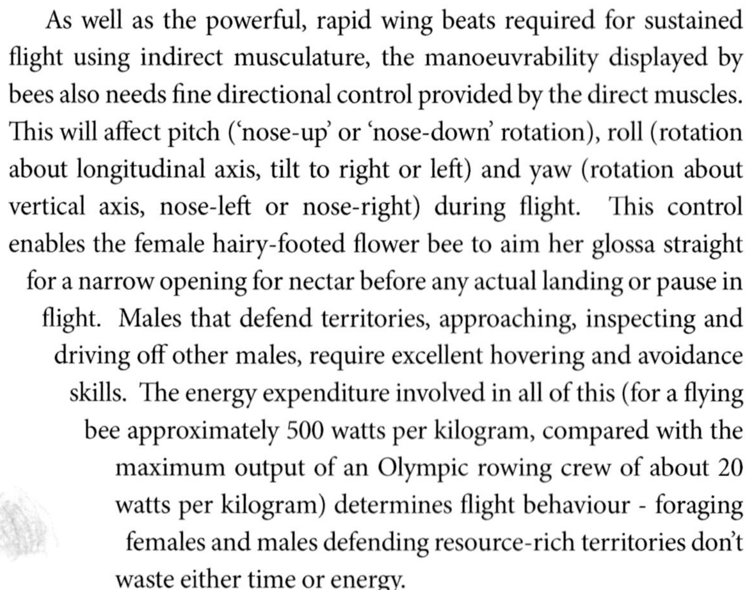

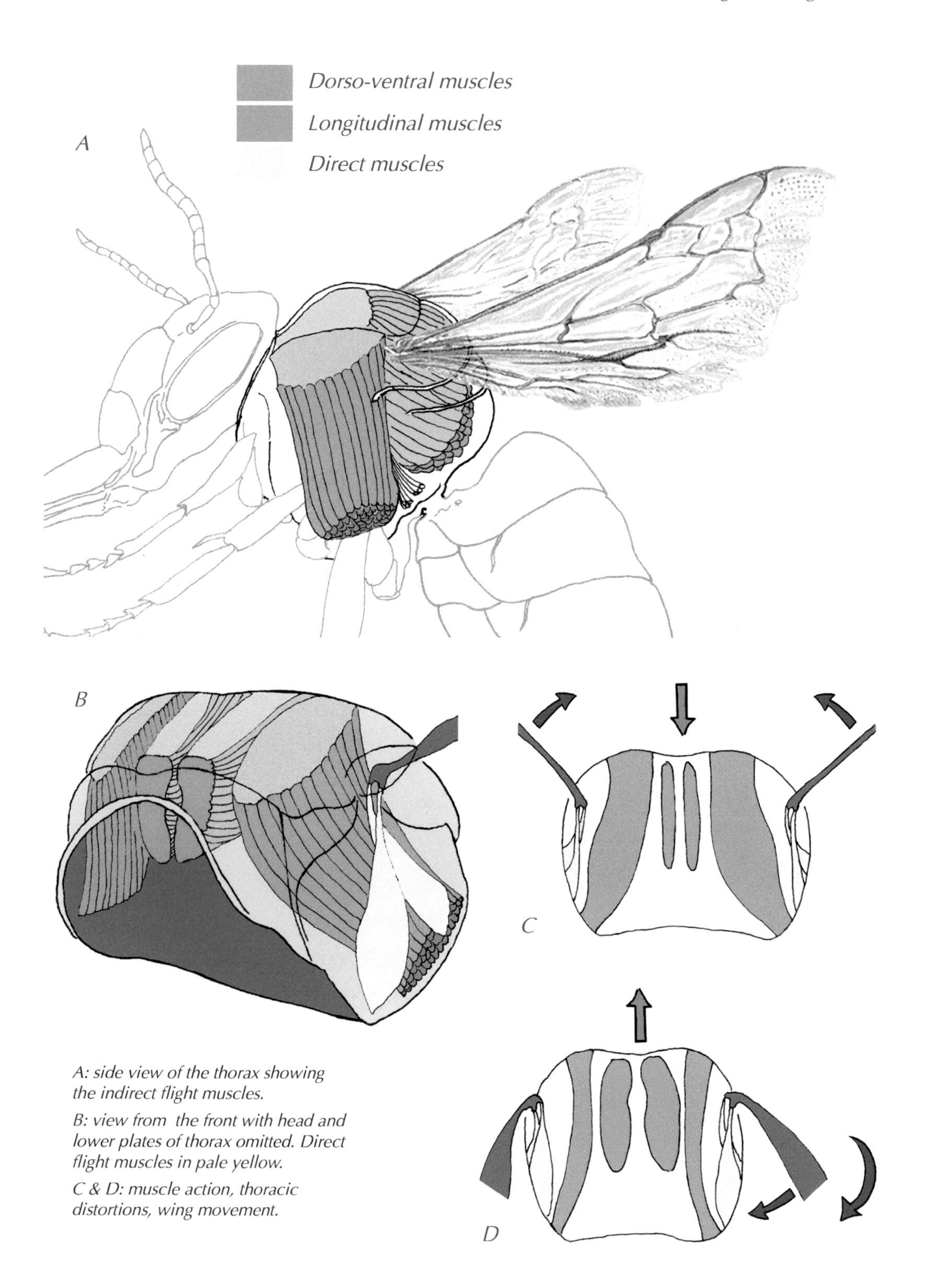

A: side view of the thorax showing the indirect flight muscles.

B: view from the front with head and lower plates of thorax omitted. Direct flight muscles in pale yellow.

C & D: muscle action, thoracic distortions, wing movement.

Wing membranes are far from simple structures. The two pairs, fore and hind, act as one during flight, with hooks (**hamuli, H**, opposite) on the leading edge of the hind wing latching on to a **coupling margin** (**CM**, below) on the trailing edge of the forewing, locking them together.

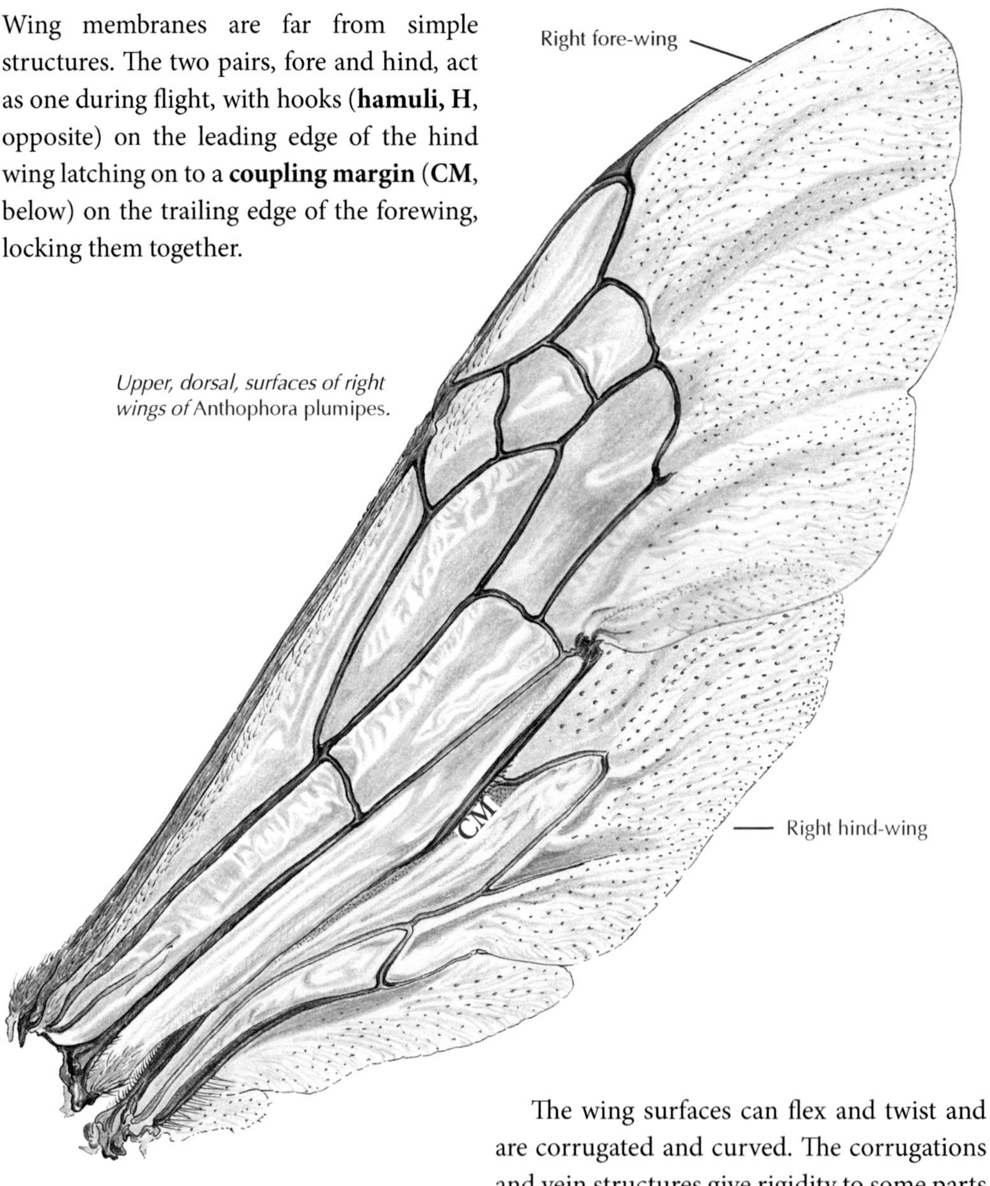

Upper, dorsal, surfaces of right wings of Anthophora plumipes.

The wing surfaces can flex and twist and are corrugated and curved. The corrugations and vein structures give rigidity to some parts of the wing, flexibility to others. There are hairs along the leading edges and small hooks across the wing surfaces. Adaptations such as these make the combined wings extremely responsive 'intelligent aerofoils' with flexible use of vortices and air flow. The bee can control the tilt and aspect of its wings, with the veins, struts and joints within the wing assisting with the rest of the fine flight control.

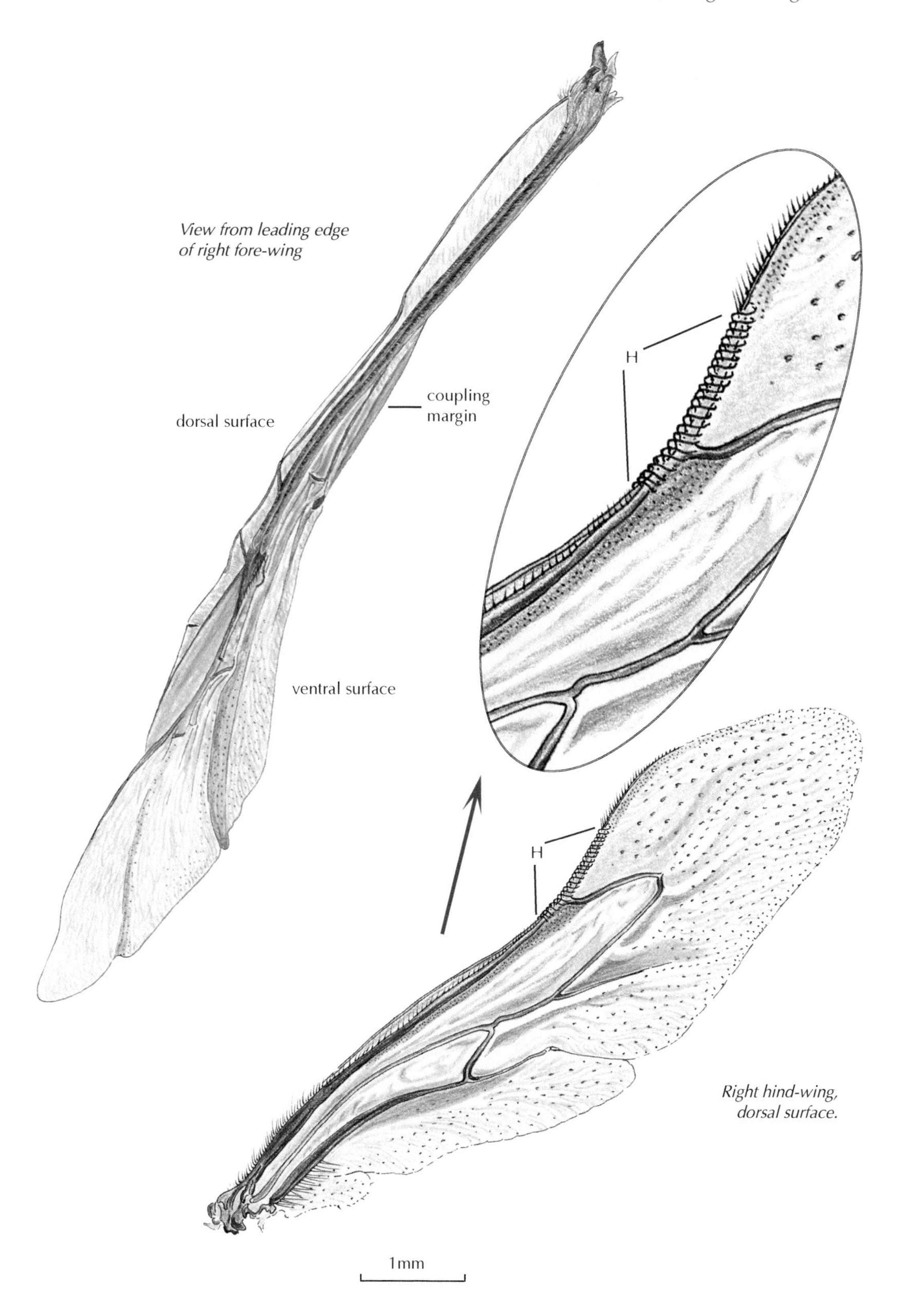
View from leading edge
of right fore-wing
coupling
margin
dorsal surface
ventral surface
H
H
Right hind-wing,
dorsal surface.
1mm

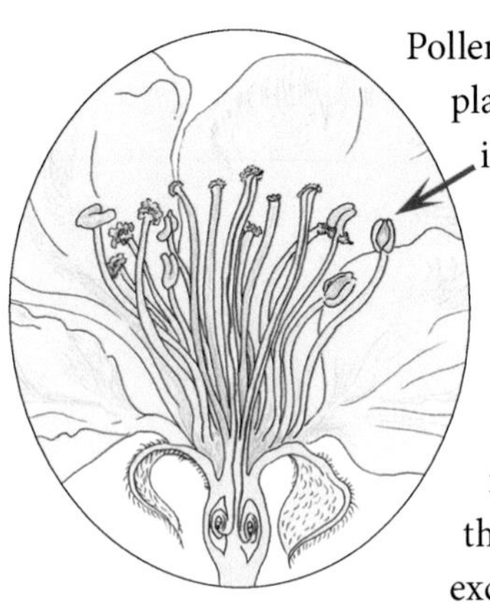

cross-section of apple flower

stigma

ovary

Pollen grains are the means by which flowering plants fertilise each other. Pollen (left) is produced in the **anthers**. When transferred from the anthers of one flower to the sticky **stigma** of another of the same species the pollen grain will germinate and produce a tube (the thin red line, right) that penetrates and grows down the **style** and into the **ovary**. Cells are exchanged and a seed can then develop. For good fruit formation, all ovules in the ovary – five in an apple and many more in a strawberry – should be fertilised, otherwise distorted fruit are produced.

Flying insects are the prime form of pollen transport put into service by many plants, and bees are amongst the most efficient of pollinators. Bees have evolved to use pollen as a source of protein and their bodies have adaptations for collecting it. The hairs of bees are crucial for carrying pollen – they are plumose (branched) to trap the grains. The hairs may be concentrated or very well developed in one part of the body to form a **scopa** or dense hairy pad. The female hairy-footed flower bee has long hairs covering most of her body and a large scopa on the tibia of her hind leg (below and right); in leaf-cutter bees, the underside of the abdomen has a thick, hairy pollen scopa (right, lower centre), as do red mason and other Megachilid bees (pp.75, 83). In some *Andrena* species there are further pollen scopae on other parts of the legs including a tuft of hairs called the **floccus** on the hind **trochanter**, and on the sides of the propodeum (e.g. in *Andrena haemorrhoa* opposite centre). Unlike *Anthophora*, however, the upper surface of the abdomen in both *A. haemorrhoa* and *Megachile*, are much more sparsely haired.

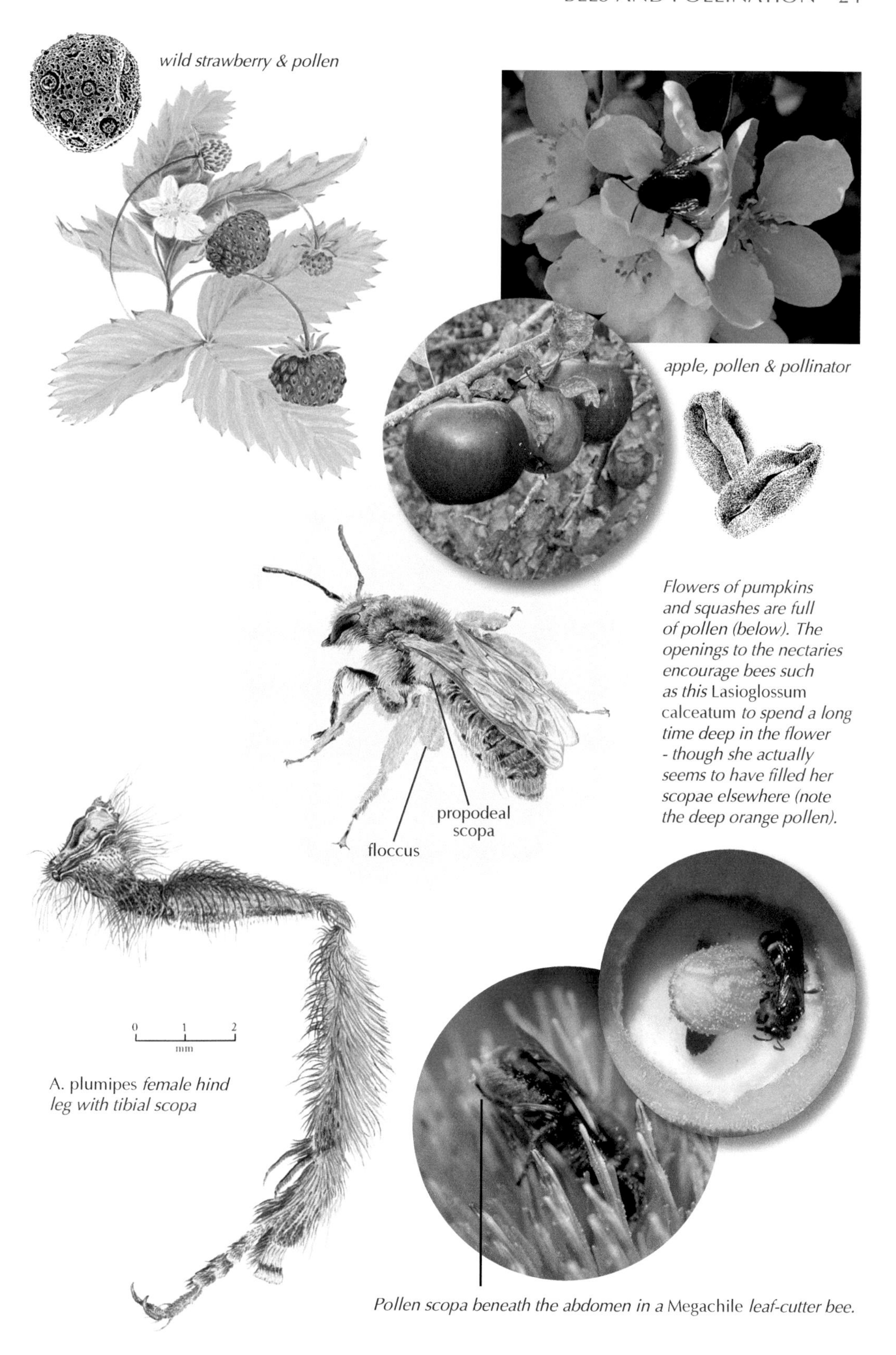

wild strawberry & pollen

apple, pollen & pollinator

Flowers of pumpkins and squashes are full of pollen (below). The openings to the nectaries encourage bees such as this Lasioglossum calceatum *to spend a long time deep in the flower - though she actually seems to have filled her scopae elsewhere (note the deep orange pollen).*

A. plumipes *female hind leg with tibial scopa*

Pollen scopa beneath the abdomen in a Megachile *leaf-cutter bee.*

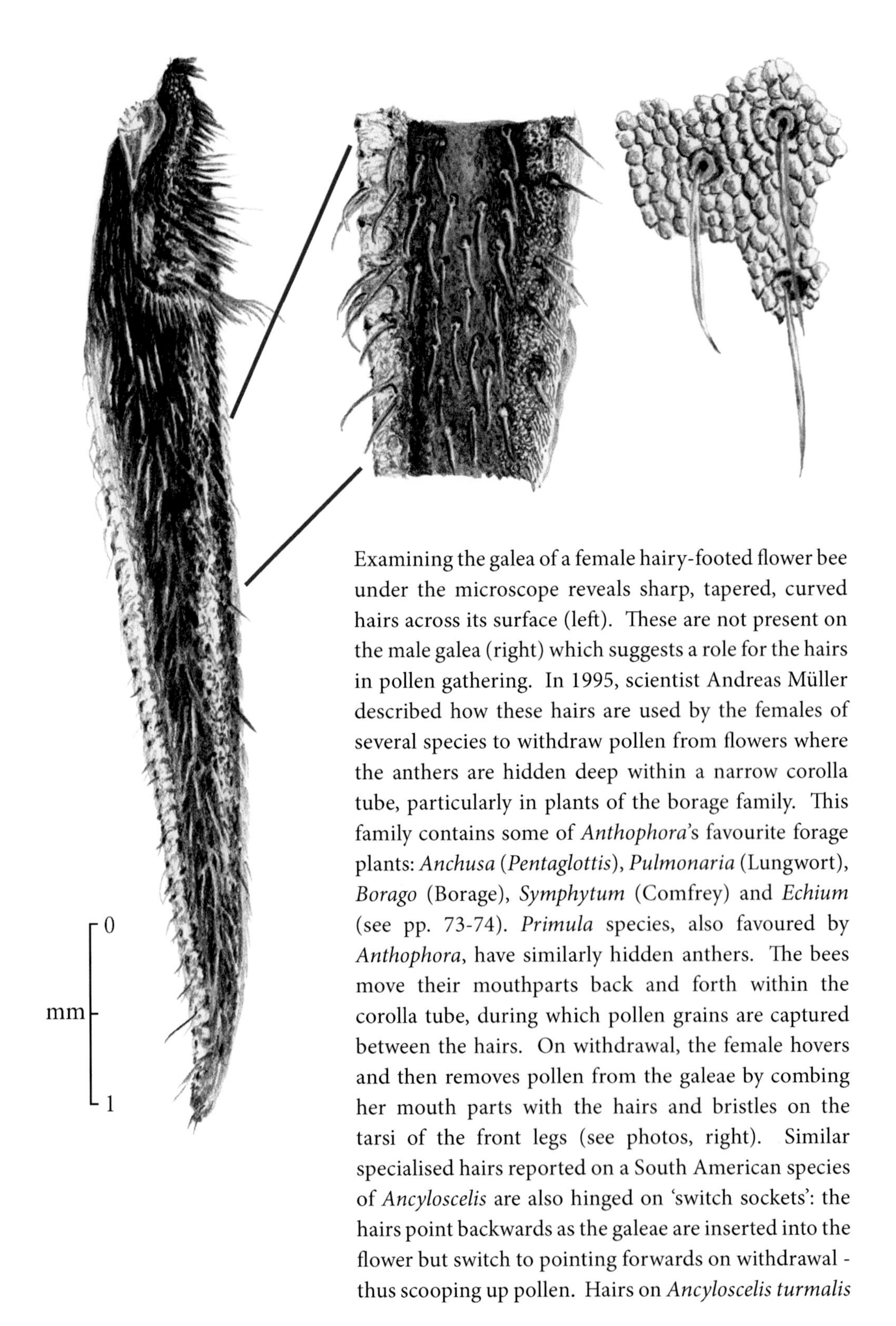

Examining the galea of a female hairy-footed flower bee under the microscope reveals sharp, tapered, curved hairs across its surface (left). These are not present on the male galea (right) which suggests a role for the hairs in pollen gathering. In 1995, scientist Andreas Müller described how these hairs are used by the females of several species to withdraw pollen from flowers where the anthers are hidden deep within a narrow corolla tube, particularly in plants of the borage family. This family contains some of *Anthophora*'s favourite forage plants: *Anchusa* (*Pentaglottis*), *Pulmonaria* (Lungwort), *Borago* (Borage), *Symphytum* (Comfrey) and *Echium* (see pp. 73-74). *Primula* species, also favoured by *Anthophora*, have similarly hidden anthers. The bees move their mouthparts back and forth within the corolla tube, during which pollen grains are captured between the hairs. On withdrawal, the female hovers and then removes pollen from the galeae by combing her mouth parts with the hairs and bristles on the tarsi of the front legs (see photos, right). Similar specialised hairs reported on a South American species of *Ancyloscelis* are also hinged on 'switch sockets': the hairs point backwards as the galeae are inserted into the flower but switch to pointing forwards on withdrawal - thus scooping up pollen. Hairs on *Ancyloscelis turmalis*

are more distinctly hooked but it is possible that the circular sockets visible (opposite, top right), enable a similar function in *A. plumipes*.

Other bee species have similarly adapted hairs, for example *Osmia pilicornis* (in which the hairs are more hooked than in *Anthophora*), *Andrena nasuta* and *Colletes nasutus*. The latter has specialised hairs on the tarsi of its front legs. 'Buzz pollination', or 'sonication', where the thorax muscles are used to shake out stubborn pollen using a high-pitched sound (p.29), is not used by these bees on borage family plants: instead they rely on the hairs. Bee species without these modified hairs on their mouthparts mostly do not take pollen from these plants: they collect nectar only.

The female, top left, has her head plunged deep into the flower of a white variety of lungwort in Greenwich Park in London. She reaches deep for the nectar whilst simultaneously collecting pollen between the hairs of the galeae.

Looking up from this patch of lungwort the flowers of a tree of the Styracaceae family are almost as busy with females. Although these flowers are quite open and do not need specialised hairs, this female is nevertheless grooming pollen from her galeae with her front legs. It will be passed back to the pollen scopa on the hind legs.

...Hairy-bodied, hairy-legged...HAIRY-MOUTHED pollen lovers!

In honeybees and bumblebees the hind legs have evolved into a complex pollen-gathering system. The 'pollen basket' – the **corbicula** – secures a mass of pollen on the tibia of the leg. Other structures include rows of spines (the **pollen brush**) and a projecting shelf (the **auricle**) on the basitarsus (part of the lower leg). The bee pushes pollen into the corbicula by flexing its leg. The grooming bee uses hairs on all its legs, some identifiable as 'combs', to move the pollen down to the corbiculae (below left). The hairs along the rim of the corbicula and a larger anchor hair in the centre retain the pollen in the corbicula. The addition of nectar makes it even more sticky. Grooming by the honeybee compresses the pollen into a tight ball enabling large quantities to be taken by thousands of worker bees back to the hive. None of this collected pollen, in all those thousands of sticky corbiculae, is available for the plant to fertilise other ovaries – a great loss to the plant.

Research on the pollination of orchards is showing that fruit set and growth is far more successful when there is a diversity of pollinator species present including social and solitary species. Increasing the number of honeybee hives is not enough. The efficiency of worker

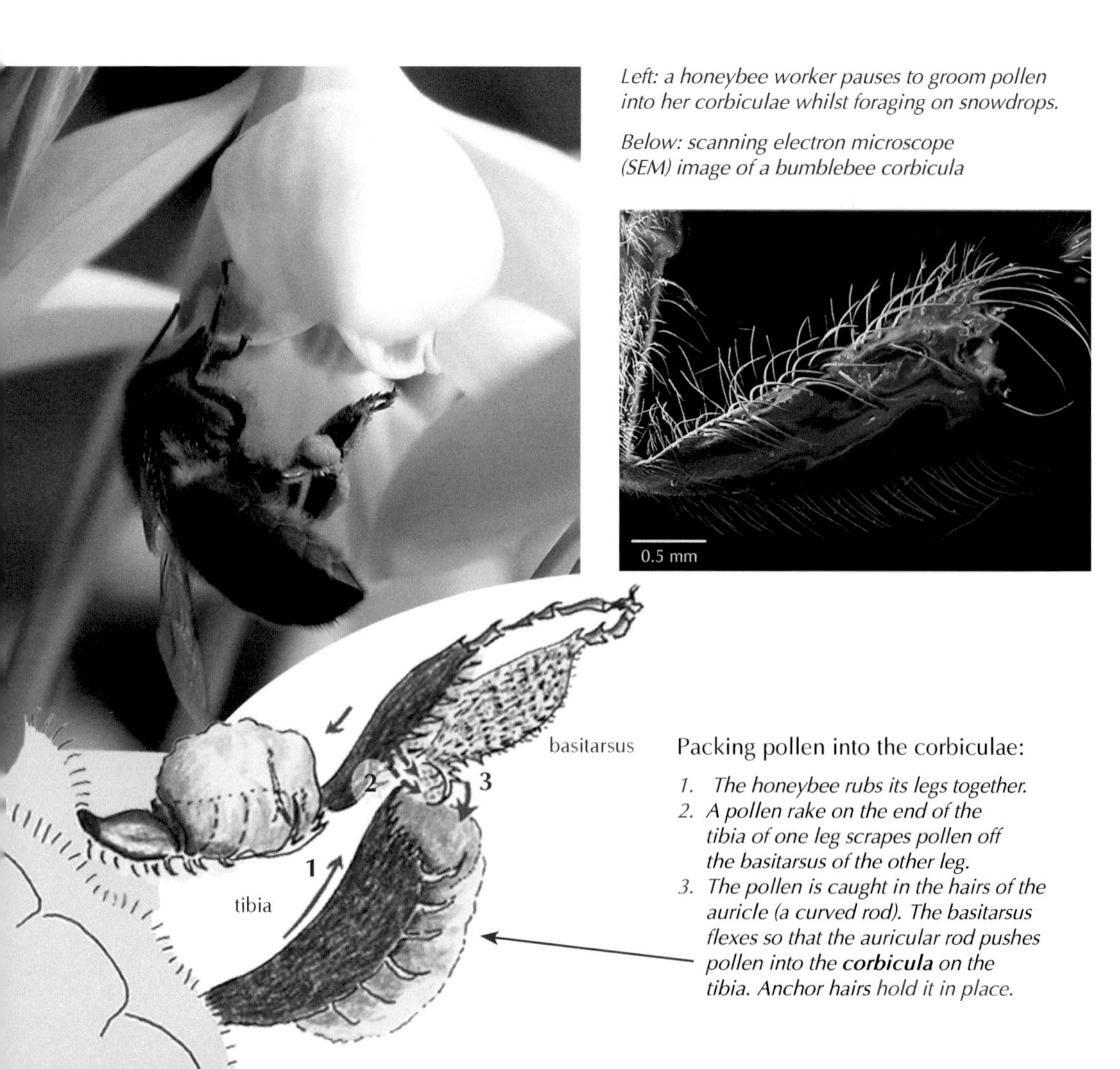

Left: a honeybee worker pauses to groom pollen into her corbiculae whilst foraging on snowdrops.

Below: scanning electron microscope (SEM) image of a bumblebee corbicula

Packing pollen into the corbiculae:

1. *The honeybee rubs its legs together.*
2. *A pollen rake on the end of the tibia of one leg scrapes pollen off the basitarsus of the other leg.*
3. *The pollen is caught in the hairs of the auricle (a curved rod). The basitarsus flexes so that the auricular rod pushes pollen into the* ***corbicula*** *on the tibia. Anchor hairs hold it in place.*

honeybees and their ability to communicate temporal and spatial variation in pollen supply to other worker bees in the colony disadvantages the orchard tree because the colony will move its foraging attention away from a crop towards other pollen or nectar sources. Thus, they will abandon apples altogether if, for example, a more sugar-rich nectar supply is available elsewhere within their range (dandelions on the orchard floor are a classic example). Honeybees tend also to concentrate their foraging effort on a row of trees, which in commercial orchards could mean between trees of the same variety. Apple varieties, however, are not generally self-fertile – they require cross-pollination from one variety to a different one in order to make fruit. Commercial orchards would have to mix a 'pollinating variety' amongst the market variety to counter this tendency. In contrast, bumble and solitary bees forage as more or less independent individuals, working intensely and rapidly across a wide area, moving constantly and visiting a wide range of different trees on one trip. A rich and diverse population of solitary and social bees in a functioning ecosystem is, therefore, most advantageous for growing human crops, solitary bees being perhaps the most important pollinator guild (a group of, not necessarily related, species exploiting the same resource).

The hairs covering the body of *Anthophora*, the much looser packing of pollen into the leg hairs, and their rapid flight during foraging at lower temperatures make them a very important early pollinator. You are far more likely to see *Anthophora* on your garden apple, pear or quince trees than honeybees (the row of espalier-trained apples above, although adjacent to two honeybee hives, is invariably pollinated by bumble and solitary bees, including *Anthophora plumipes, Andrena haemorrhoa* (above, right) and *Osmia bicornis*). Males also carry pollen (inset) during feverish feeding flights in their continuous search for mates (unlike male honeybees). The production of pollen is expensive for plants – from their view of the pollination arms race the social honeybee has evolved highly 'selfish' means of collection whereas solitary bees are more 'magnanimous'.

With *Anthophora*, therefore, we have a....

'first class flower bee pollen post!'

Buzz pollination, or **sonication**, is a further complication in the pollen arms race. To make the distribution of pollen more efficient for the plant – to avoid giving it away too generously to pollen eaters – the anthers of some groups have become almost closed tubes with the only exit for pollen being through a narrow slit or even a tiny pore. Unable to simply brush against the anther, bees such as bumblebees and some solitary bees, including *Anthophora* species (but not honeybees), have evolved the ability to use their indirect wing muscles, without moving their wings, to create a sound that is higher pitched than their usual 'buzz' – hence 'sonication'. The vibration causes the pollen in the anther to move rapidly, increasing its electrical charge and eventually causing an almost explosive escape from the pore. The bee positions itself beneath the anthers, forming a hairy cup to receive the shooting pollen. The flower forms of such plants, for example in the family Solanaceae (below), encourage this positioning by the bee. The plant therefore releases pollen only to certain pollinators. Because they have hairs all over their bodies the targeted bees are also good at passing on pollen to the next flower. Up to 20,000 plant species are adapted in this way, including many important food crops such as tomato, sweet pepper and aubergine (all members of Solanaceae). Agricultural systems must take this into account in their design if they are to ensure good pollination of such crops.

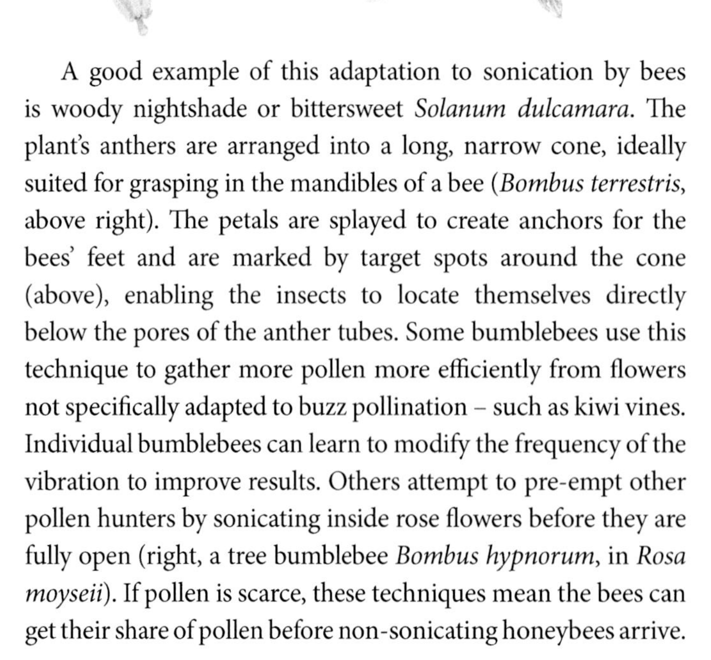

A good example of this adaptation to sonication by bees is woody nightshade or bittersweet *Solanum dulcamara*. The plant's anthers are arranged into a long, narrow cone, ideally suited for grasping in the mandibles of a bee (*Bombus terrestris*, above right). The petals are splayed to create anchors for the bees' feet and are marked by target spots around the cone (above), enabling the insects to locate themselves directly below the pores of the anther tubes. Some bumblebees use this technique to gather more pollen more efficiently from flowers not specifically adapted to buzz pollination – such as kiwi vines. Individual bumblebees can learn to modify the frequency of the vibration to improve results. Others attempt to pre-empt other pollen hunters by sonicating inside rose flowers before they are fully open (right, a tree bumblebee *Bombus hypnorum*, in *Rosa moyseii*). If pollen is scarce, these techniques mean the bees can get their share of pollen before non-sonicating honeybees arrive.

Time and motion: if you watch foraging bees of many species you will see that they often approach a flower but then back off without landing. Pheromones from the tarsal glands (p.12) of honeybees leave a trace on the flower. This draws attention to the food source and indicates when a flower has already been harvested. Honeybees have evolved highly efficient techniques for exploiting their environment but this is important for all bee species – females cannot afford to waste energy landing and taking off from a flower whose rewards are already depleted. Bumblebees are similarly sensitive to scent marks on flowers and avoid those previously visited by themselves, others of their own species or bumblebees of other species. Given the chemical similarities of many bee pheromones it is possible that bees are sensitive enough to their chemical environment to avoid depleted flowers visited by a range of other bee species. These scent markings are relatively short-lived: once the plant has recharged its nectaries, bees will re-visit.

Electro-reception in bees: plants, their flowers and pollen, being earthed, tend to have a natural negative charge. Flying insects automatically generate a positive charge whilst moving through air molecules. When a bee approaches a pollen-laden flower any disturbance to the pollen will cause it to be attracted to the bee's body. This helps the bee greatly – for example during the 'shovelling' motion of the leaf-cutter bee moving pollen into its abdominal scopa (p.24). A visit by a positively charged bee will then alter the electrical field of a flower. If a bee could assess the shape and intensity of the electrical field as it approaches, it could avoid landing on flowers with modified fields. There is now some evidence that bumblebees are able to detect the intensity and shape of the electrical field of a flower. In addition to visual and odour cues, this ability may have influenced the evolution of mutually beneficial communications between plant and pollinator.

Pollination deception: the early spider orchid *Ophrys sphegodes* is pollinated by males of the common buffish mining bee *Andrena nigroaenea* (above, a female foraging on dogwood). The orchid comes into flower to coincide with the early spring flying of the males. The sex pheromone of the female *Andrena* is imitated closely enough by the orchid flower to fool male bees into attempting copulation, during which pollen packs (**pollinia**) are attached firmly to the male bee. However, warmer springs caused by climate change may be bringing forward the emergence of the males. The flowering time of the orchid is not advancing so quickly and pollination could suffer as a result. Unfortunately, the early spider orchid is a rare chalk downland specialist and is unlikely to be found in our gardens (unlike its bee pollinator).

A female red mason bee *Osmia bicornis* foraging on barrenwort *Epimedium* flowers. This sequence shows how the plant obliges the bee to hold its body. The nectar is stored as pools in the brownish spurs at the back of the small yellow petals, surrounded by the large yellow sepals. The bee can land below the flower, clutching the four anthers and the central stigma between its legs. Its abdomen naturally arcs beneath the stigma and anthers. The proboscis has to enter between each petal and the base of the style. The glossa (visible, below left) has to reach down the spur to the nectar. This female was foraging primarily for nectar, not pollen: the plant is thus the main beneficiary of any incidental pollen transport.

Barrenwort is a very popular garden plant for shaded and semi-shaded areas such as under tree canopies. It is found from North Africa to Japan and has greatest diversity in China. It overlaps with the red mason bee in North Africa and the Eastern Mediterranean.

Flowers mature progressively along the inflorescence. On each flower, the stigma is receptive to pollen at a different time to the anthers producing the pollen. When the bee is foraging on the flower the stigma presses against the scopa on the underside of her abdomen, picking up any pollen from another flower. Barrenwort is not self-fertile, so pollen is needed to cross from one plant (or clone) to another: a travelling, nest-building *Osmia* is perfect. The red mason bee is a generalist feeder, though, and visits many different flowers – this female went from barrenwort flowers to green alkanet to rock cranesbill (*Geranium macrorrhizum*) in rapid succession. In the native range of barrenwort, *Osmia* or *Bombus*, with their rounded, hairy abdomens, would seem better pollination partners for the flower than smaller *Andrena* species, whose females efficiently gather pollen on their hind leg (their males are probably more efficient pollinators than the females).

The family Lamiaceae (sometimes called Labiatae) includes some beautiful native plants such as white deadnettle *Lamium album*, left, all the mints, thymes, sages and lavenders. All are great forage plants for long-tongued bees. Another is Turkish sage *Phlomis russeliana*. Many Lamiaceae have striking hooded flowers with deep tubes of fused petal bases requiring a long glossa. The anthers are held just below the upper petal hood so that when a bee reaches for nectar, whilst resting on the lower petal lip, pollen is rubbed onto the head or hairs of the thorax. The stigma is longer than the stamens and when ripe can reach down to receive pollen from the bee's abdomen hairs. A worker *Bombus pascuorum* prepares to enter the *Phlomis* flower, left and below, head pushing under the anthers, antennae drawn back. Once inside the flower the weight of the bee causes the stamens to brush pollen onto its abdomen. This bee already has pollen packed into her corbiculae, and one of the plant's anthers may even have reached directly to the corbicula of her left hind leg (not necessarily in the plant's best interests). As long as the bee does not groom the pollen completely from its body hairs before visiting the next flower, the *Phlomis* has ample opportunities to be fully pollinated.

Later, a female *Megachile ligniseca* approaches the same patch of *Phlomis* (above). She hovers, weighing up the flower and how to enter it. Evidently interested in pollen as much if not more than nectar, she niftily inverts herself just before entering. Clinging to the flower's stamens and stigma with her legs, she is then in a similar position to the *Osmia* on barrenwort (p.31). The anthers brush pollen directly into her pollen scopa under the abdomen.

Both plant and bee benefit from this ability of the leaf-cutter female to analyse and then devise a manoeuvre to suit the morphology of the flower. In this situation she does not have as much need to use the shovelling action with her hind legs that she often employs to loosen and direct pollen into the scopa on other flowers (p.83).

When visiting a sweet pea, a female leaf-cutter engages her legs extremely actively and accurately. The images on the right are stills from a video. They are blurred because the bee's action is so vigorous. The plant's stamens and stigma have been forced up out of the keel – the lower petals that surround them – by the weight of the bee. Not content with the pollen coming into contact with her scopa, she scrapes the stamens and anthers towards the scopa. This takes just seconds and she is not on the flower for long.

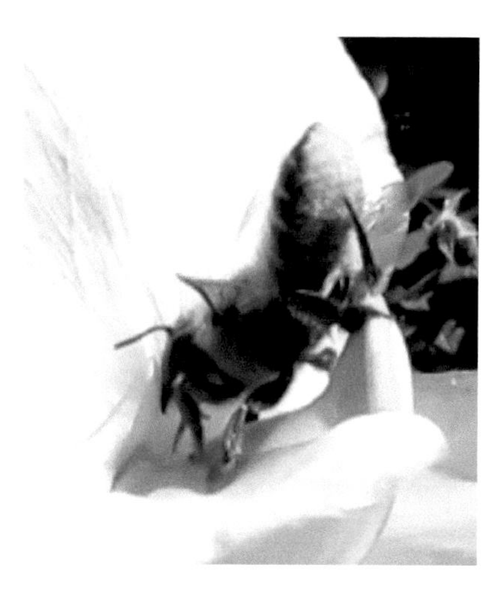

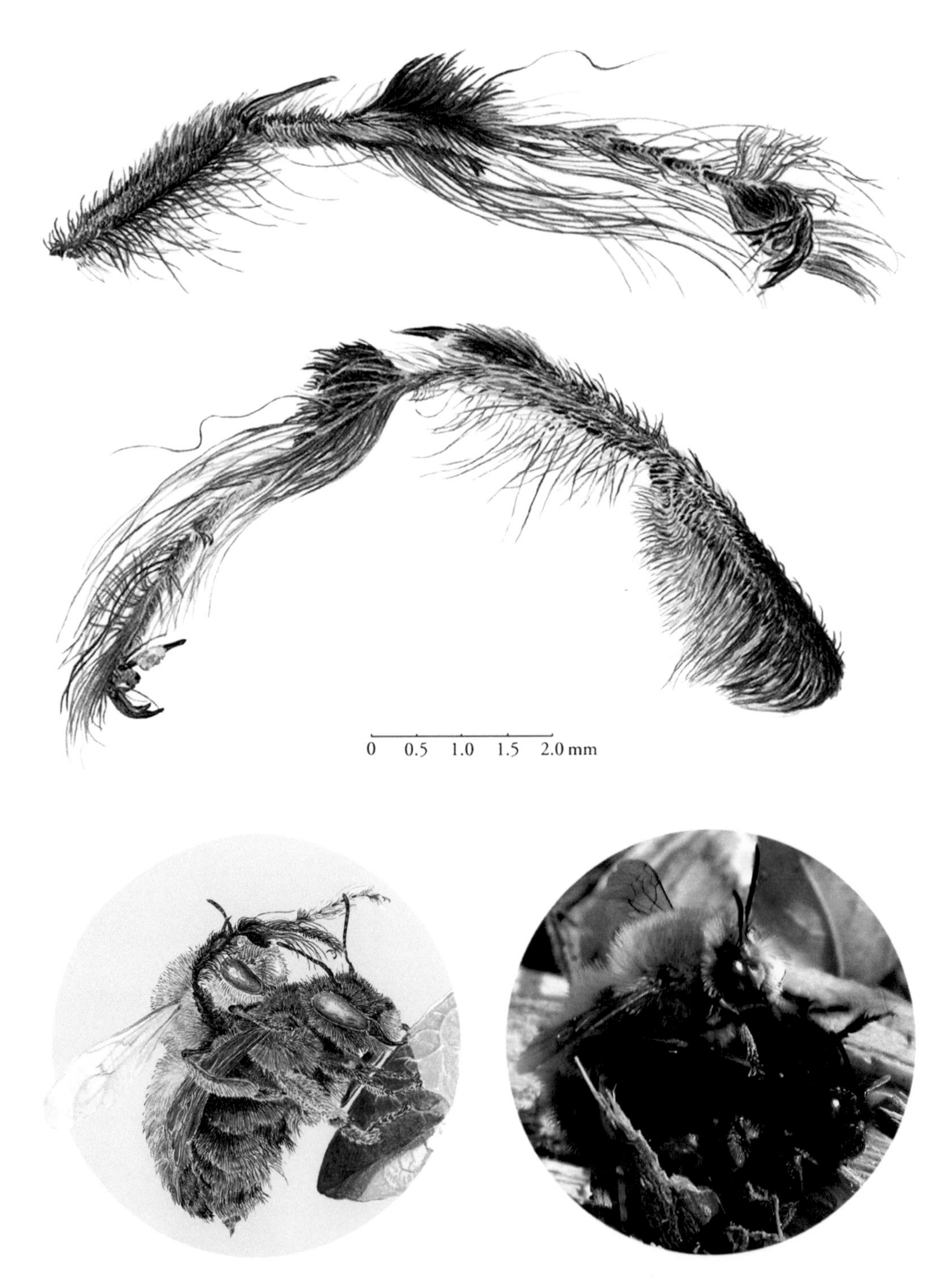

The beautiful long hairs and striking fans on the long, middle legs of the male Anthophora plumipes *(top and centre) must have a particular function. During mating the male holds his mid legs above the female's head (above, left). Males of most solitary bees emerge before females. They are waiting to mate with the females as soon as they emerge from nest holes. Mating has to take place quickly and efficiently – the weather could turn against them. Males must be ready.*

Females are also ready to mate a short time after they emerge. Once mated, however, their focus is on building and stocking nests. Further males attempting to mate will get in the way – the female must discourage them.

When searching for a mate the male hairy-footed flower bee patrols suitable food plants and hovers close by foraging females. Several males will come as close as possible, trying to assess whether the female has been mated already and watching the other males (larger males have more successful matings in these situations). Whilst hovering, males transfer a sex pheromone (a secretion) from an abdominal gland onto the hairs of the middle legs before approaching the female (above).

Once he is in mating position, grasping the female with fore and hind legs (opposite, lower left), he will waft the pheromone by flicking each mid leg alternately across the tips of her antennae. This may pacify her and stop her resisting during the lengthy mating. The raised legs may also fend off other males. Females that have been mated and have this pheromone still present are not usually approached, avoiding conflict and wasted energy. If a female is bothered by males when she is trying to build a nest, she will forage on flowers part hidden by leaves, where she is less visible and less accessible (as the female is doing above amongst white deadnettle) or, if necessary, she will land and crouch, tucking in the abdomen and raising a foreleg to fend off the troublesome male (opposite, lower right). It is interesting to note that the sound of the flying male – its buzz – is of a different pitch to that of the female. This can alert human observers to the male behaviour, but whether the female, who would want to know they are approaching, can hear this is not known.

There is a clasping and a stroking, a flicking and a fanning of pheromone.

Bumblebees are very hairy all over.

Flower bees are also hairy all over, especially their legs.

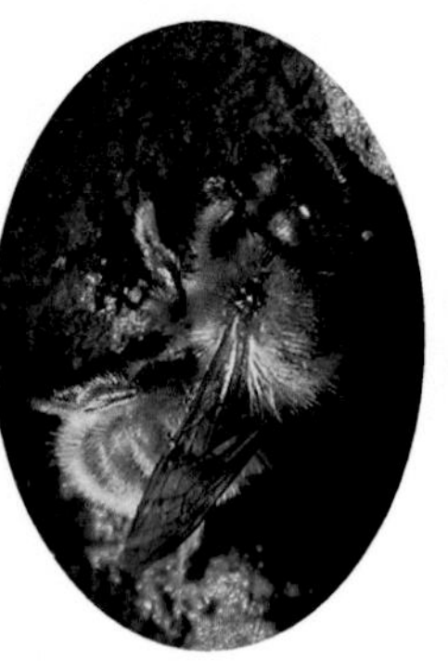

Red mason bees have a hairy abdomen and a hairy face.

Leaf-cutter bees have a particularly hairy belly (beneath the abdomen).

We have seen how the hairs of the hairy-footed flower bee act as insulators, pollen collectors and pheromone fans : bee hairs have evolved to fulfil many different purposes.

The hairs on the legs are stiffened and of different lengths for pollen collection, in grooves or shaped as brushes to facilitate grooming; 'combs' enable the removal of pollen from the body into the nest chamber. Many bees that dig their own nest holes – *Anthophora* and other mining bees – have strong spines and spurs on their legs to help the bee brace itself against the nest tunnel and to help loosen and excavate grit and soil. The female *Anthophora* hind leg has a pollen scopa, spines and combs (left).

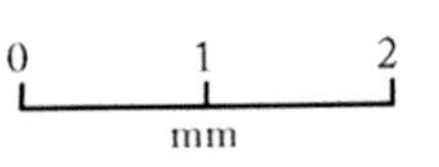

Many hairs are movement sensors – they are **mechanoreceptive** – and have nerve endings at their base that provide important sensory information about which way is up, body position, key bodily movements such as the functions and positions of legs and wings, and wind direction and strength.

Other hairs act as taste sensors – contact **chemoreceptors** – on the mouthparts, antennae, feet and other parts of the body. The mechanoreceptor hairs at the two main joints of the antennae help the bee control the direction of antennal movement (opposite, upper). Those near the tip of the antenna are mechano- and chemoreceptor hairs (opposite, lower).

Above: Yellow-faced bees are tiny and have hardly any hairs.

Above right: Honey bees are less hairy than others but they do have hairy eyes! In the SEM image, above right, the compound eye is visible behind the antenna).

*Above (far) right and below: the honeybee antenna has mechanoreceptor hairs on the pedicel (***P***) where it hinges with the scape (***S***). When the front parts of the antenna (the pedicel and the flagellum,* **F***) are moved up and down these hairs are pressed against the scape. This enables sensitive control of the antenna, but also provides information on flight speed when the antenna is deflected by wind.*

When bees groom they use a series of combs and bristles on their legs. The front legs also have a neat way of cleaning the antennae. Sense of smell is very important to bees and the antennae can pick up the slightest of scents – but they must be clean. The sensitive hairs and nerve pits visible on the SEM images can get clogged with debris. A circular cleaning notch on the front legs of bees fits exactly round the antenna, and has tiny bristles that clean off the finest dust. A small rod (the **fibula**) on the tibia next to it holds the antenna firmly in place whilst the bee drags her leg down the length of the antenna, very efficiently clearing out debris.

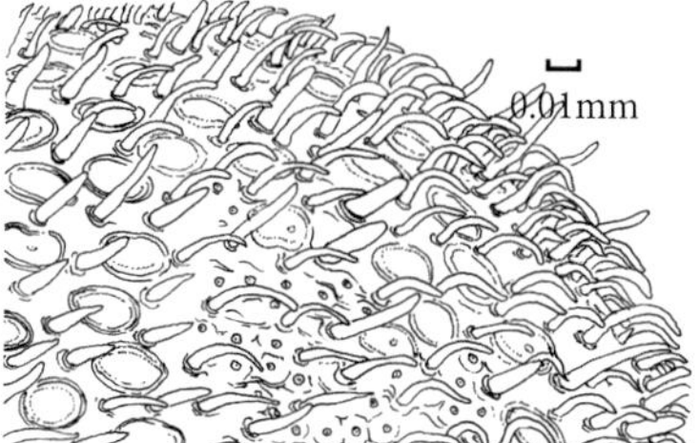

Chemoreceptor hairs on the tip of a honeybee antenna

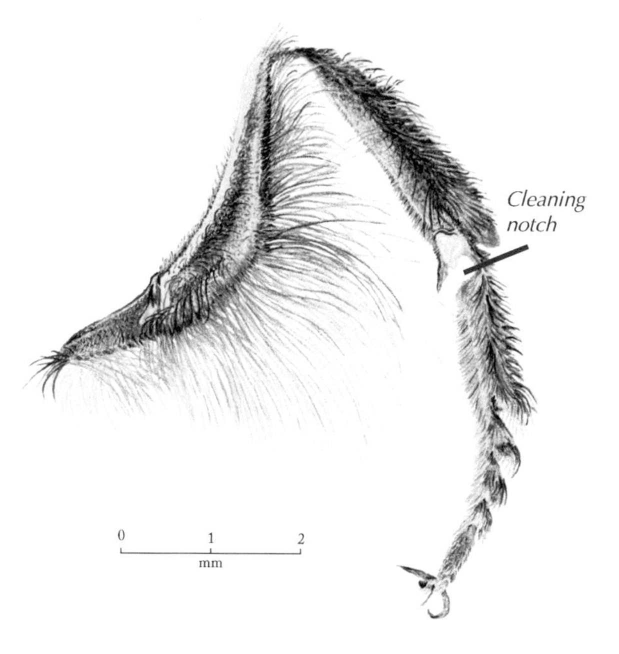

Nectar is a fluid produced by plants to lure flying pollinating insects. It is rich in a range of sugars, primarily sucrose though in some plants there will be glucose and fructose. Sugar concentration is generally 35–45% (but ranges from 7–70%). Nectar also contains small but important quantities of minerals, vitamins, organic acids and aromatic compounds. Some plants use caffeine in nectar to stimulate pollinators' memories, ensuring their return. Nectar is exuded from the nectaries, which are usually at the base of a flower so that the insect is drawn down past the pollen-bearing anthers.

The mouthparts of bees

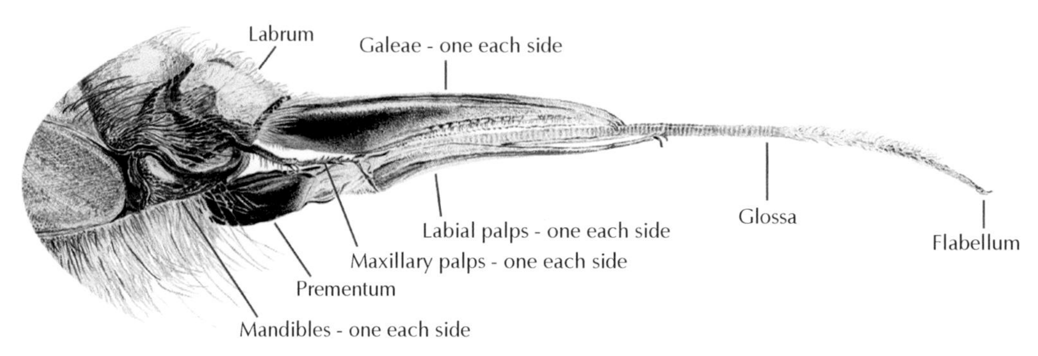

The galeae and labial palps protect the bee's glossa and create an effective tube around it for collecting nectar. Nectar sticks to the hairs on the glossa and the spoon-like tip of the **flabellum**. Its many hairs enable capillary action to draw nectar between them and they expand the size of the glossa by up to 50%. They are also chemoreceptive hairs for detecting the nature of the nectar. The relative movements of the various mouthparts – the prementum, galeae, palps and glossa – bring the nectar up into the head in a cycle of loading, retraction and unloading. The part of the glossa projecting beyond the galeae is flexible and mobile. Watch *Anthophora* approaching a flower to feed and you will see the glossa exploring and preparing to enter and travel deep into the flower.

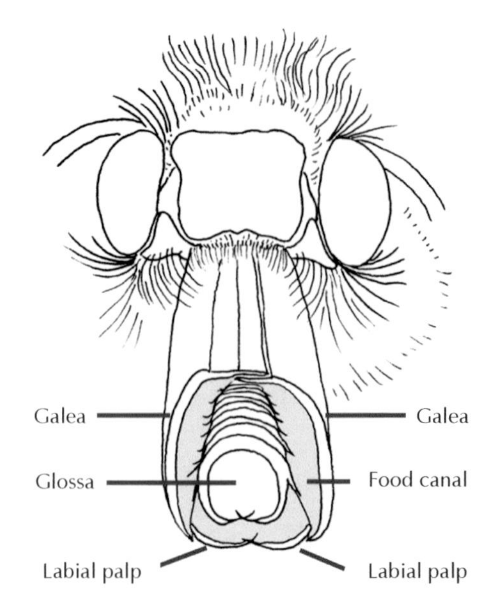

Solitary bees can refine nectar by 'blowing bubbles' to evaporate water.

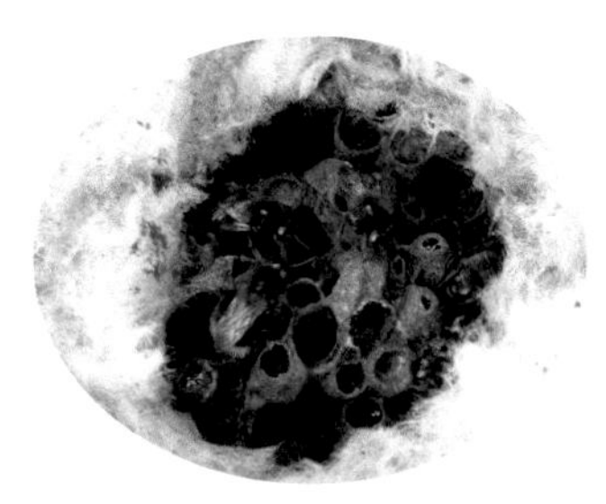

Bumblebees make honey in small pots.

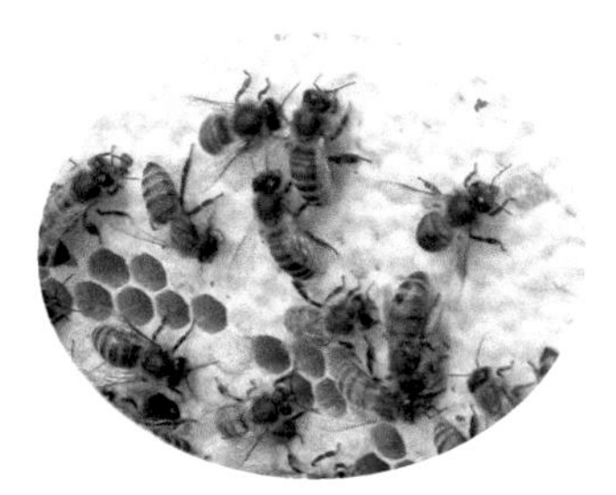

Sophisticated honeybees make loads of long-lasting anti-bacterial honey which is stored in their honeycomb.

Nectar is used directly as a source of energy (carbohydrate). Solitary bees mix it with pollen to make a rich 'cake' in the nest chamber for the growing larvae. Bumble and honeybees also feed it to larvae, but both store it separately from pollen in honey pots or honeycomb. However, nectar would easily grow bacteria and mould if the bee tried to store it straight from the flower, but if the mother bee can reduce the water content and raise the proportion of sugars, the nectar can be stored more safely for the young. If she also adds enzymes from her hypopharyngeal glands, sucrose in the nectar will be converted into fructose and glucose. These sugars are more soluble than sucrose and can increase the sugar concentration to over 80%. This refining of the nectar along with water evaporation makes the nectar a more valuable food that is resistant to bacteria and mould.

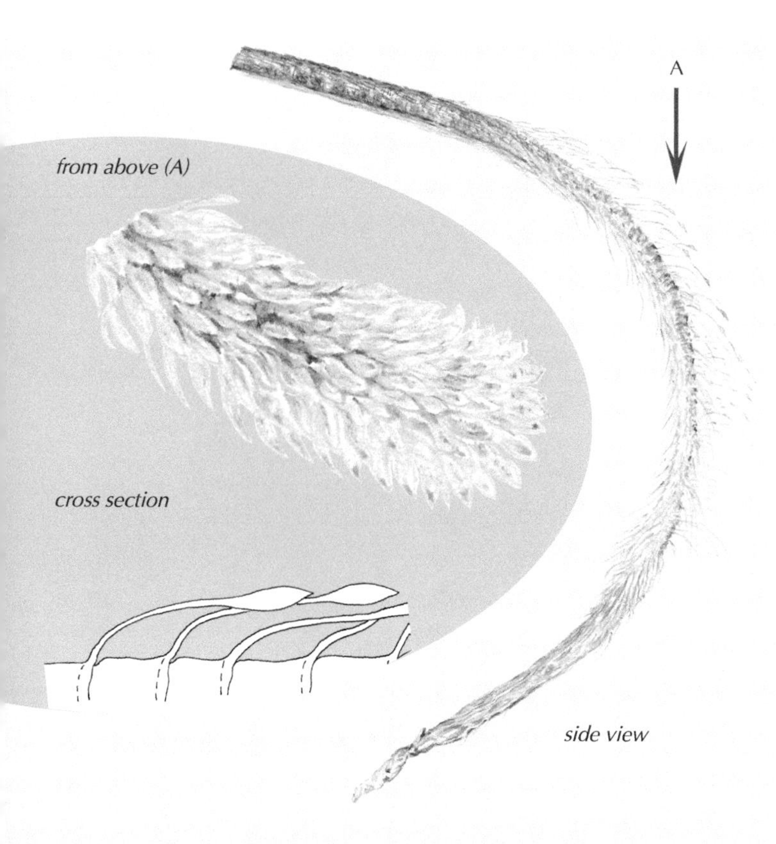

Left: close-up of the tip of Anthophora*'s glossa and flabellum showing the many tiny hairs that help to draw up nectar by flexing and filling with the sticky fluid.*

Inset: hairs widen at their tips into a spatula form, improving nectar uptake.

Lower inset: diagrammatic cross-section showing rings of cuticle that make up the glossa, with strengthened rims from which the hairs arise. Between these the cuticle is membranous for flexibility and expansion.

*Tube-shaped flowers - you need a very long tongue (*Anthophora *and cowslip).*

Bell-shaped flowers - you need a moderately long tongue (honeybee and Campanula*).*

Saucer-shaped flowers - you can have a short tongue (fly and buttercup).

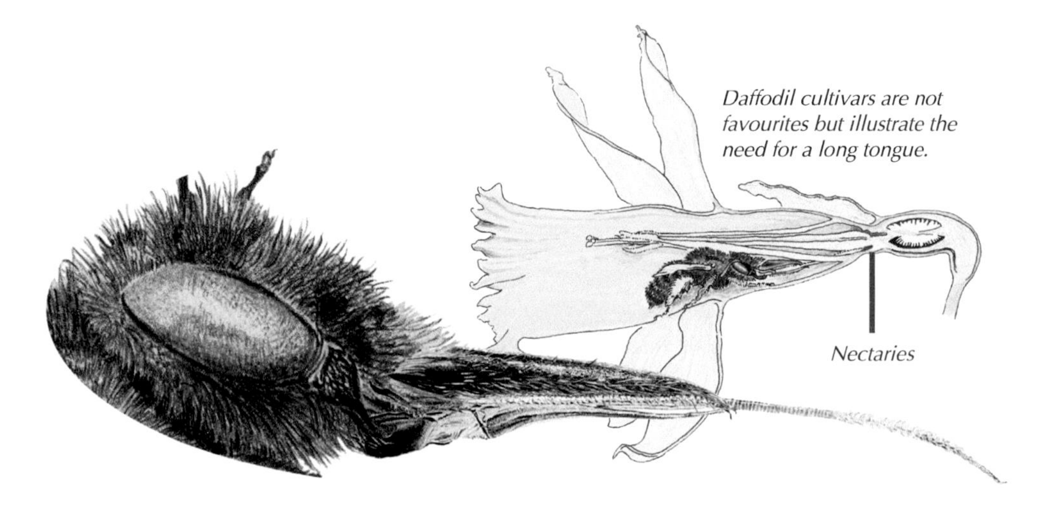

Daffodil cultivars are not favourites but illustrate the need for a long tongue.

Anthophora plumipes is a generalist feeder – it will use a wide range of plants for nectar and pollen (it is **polylectic**). When it is flying in early spring, many of its food plants have tubular flowers or flowers with nectar set deep within. These include cowslip *Primula veris, Geranium macrorrhizum*, grape hyacinth (*Muscari* species) and wild daffodil *Narcissus pseudonarcissus*. Other favourites with deep nectar tubes are lungwort and comfrey *Symphytum*.

You might sometimes see nectar seeping from the undersides of leaves or from glands on leaf stems (Passion flower Passiflora caerulea*).*

Even when folded away *Anthophora's* mouthparts stretch underneath the bee almost to the abdomen. When reaching for nectar the mouthparts extend to 13mm; the whole body is only 14mm long.

Other bees also have very long tongues. Left, the garden bumblebee *Bombus hortorum* on a *Phlomis* flower; below, the blue mason bee *Osmia caerulescens* is 8mm from head to tail. Its mouthparts stretch to 6.5mm.

The many species of solitary bee have lifecycles that are variations on the pattern shown here by a species of *Megachile* Leaf-cutter bee. These variations are explored in 'Other bees to spot' 1-14, between pages 71 & 110.

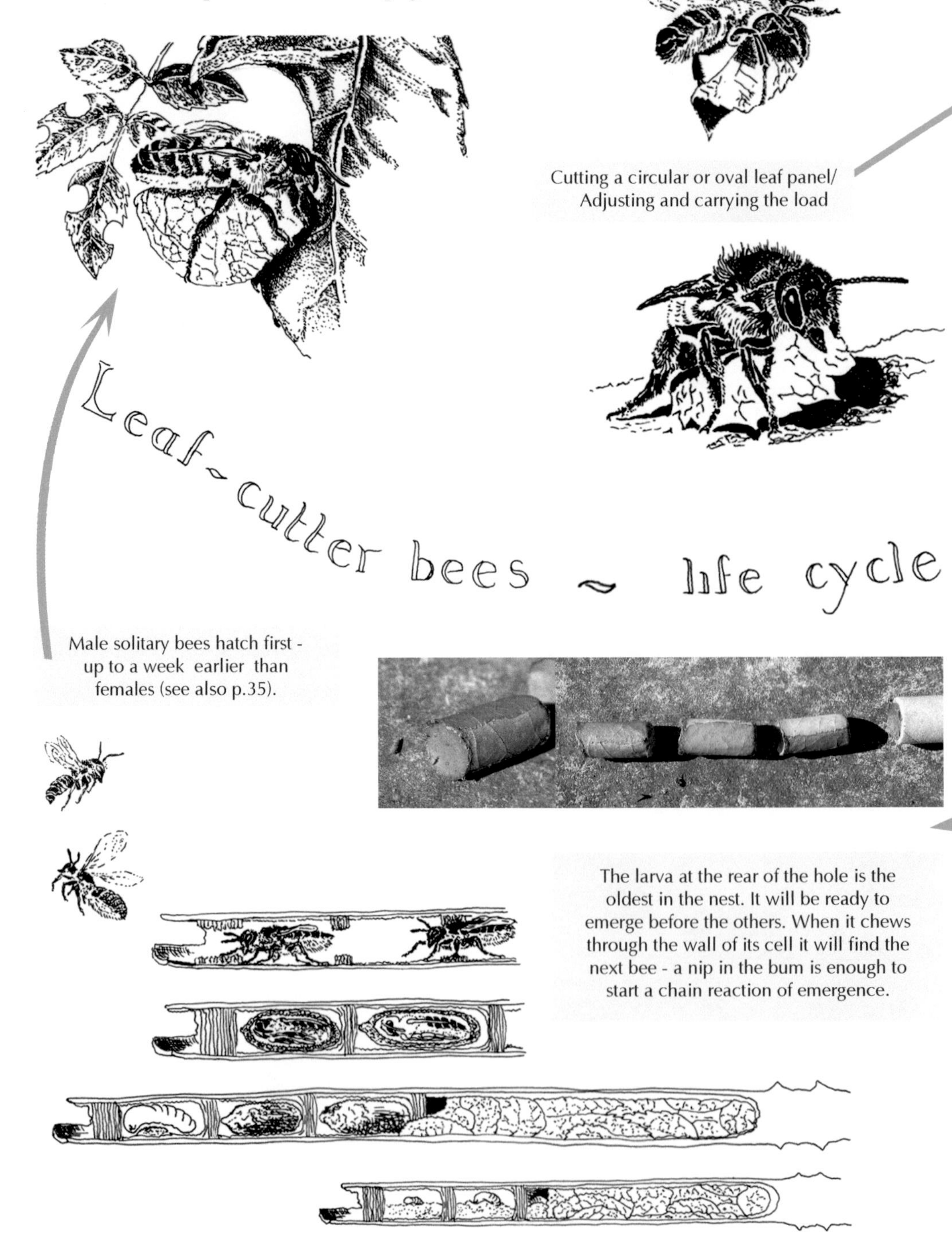

Cutting a circular or oval leaf panel/
Adjusting and carrying the load

Male solitary bees hatch first - up to a week earlier than females (see also p.35).

The larva at the rear of the hole is the oldest in the nest. It will be ready to emerge before the others. When it chews through the wall of its cell it will find the next bee - a nip in the bum is enough to start a chain reaction of emergence.

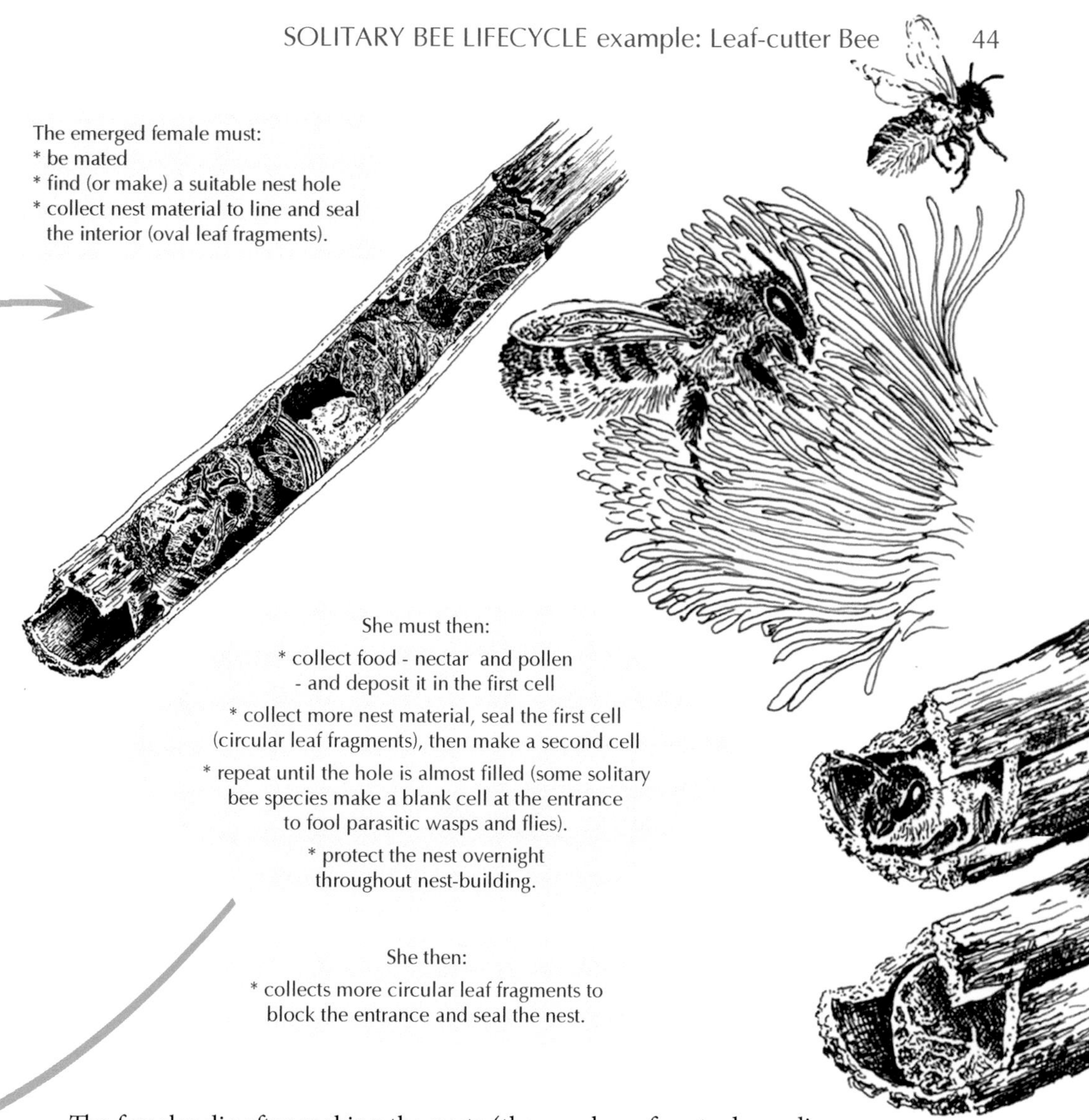

The females die after making the nests (the number of nests depending on suitable weather and habitat).

Inside the cells the eggs hatch and the larvae grow on the food left by the female.

The young bee over-winters in the hole as a larva or as a pre-pupa, or sometimes as a dormant adult, depending on the species.

Other species use different nest materials, such as mud, sand and grit, the hairs from the surfaces of leaves, tree sap and other plant resins, chewed up leaf fragments mixed with a sticky glandular secretion, or the secretions alone.

For nest sites, solitary bee species use old beetle larvae holes, cracks and crevices in wood, hollow sticks, bramble stems or reeds (including on thatched roofs), holes in stone or brick walls, even old snail shells, as well as in holes provided deliberately or accidentally by humans. Mining bees (such as *Anthophora*) make their own holes in the ground in bare soil, in sandbanks, soft coastal cliffs or in the old mortar of walls. Many mining bees line the holes with secretions that waterproof the hole and prevent the growth of fungi and bacteria.

In the soft mortar holding together the stones of an old garden wall a female is digging a new nest hole. Another female has finished a cell in her nest chamber deeper in the wall. She has laid an egg on top of the food she has left there and sealed the first cell. This will have taken most of a day. In the second cell she has laid another egg. She may add another cell before sealing the tunnel with a concave plug of clay and then moving on to make another tunnel. A nest may take several days to complete.

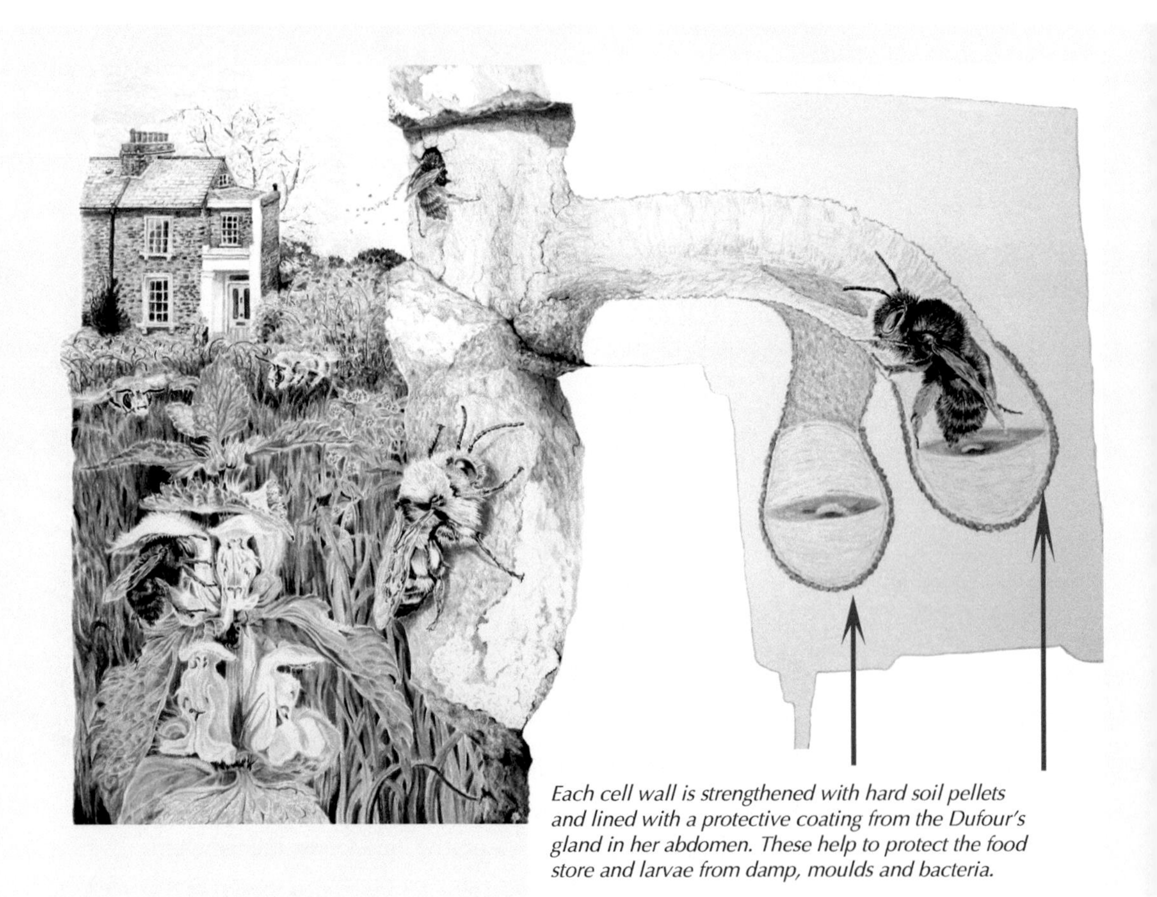

Each cell wall is strengthened with hard soil pellets and lined with a protective coating from the Dufour's gland in her abdomen. These help to protect the food store and larvae from damp, moulds and bacteria.

On a warm stone outside the hole of the female *Anthophora plumipes*, a female cuckoo bee *Melecta albifrons* waits. She will try to sneak into the nest to lay her own egg. This will hatch before *Anthophora's* and the cuckoo larva will take over the cell in which it has been born. Below the wall (left), a female flower bee finds nectar deep in the flower of a white deadnettle. Two males hover, hoping to mate, even though she is already building a nest in the old wall. Beyond a bramble patch, a town house* has nest sites in the mortar of its chimney and a flowering cherry is covered with feeding females. For more on cuckoo bees and *Melecta*, see pp.107-108.

* This book was designed behind the tiny window in the gable end.

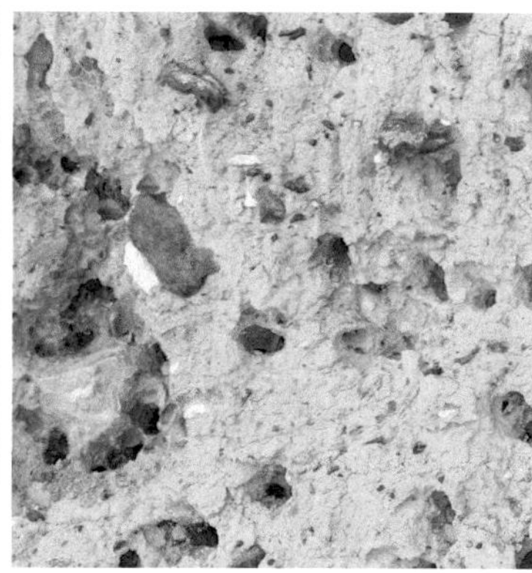

These walls are made of cob – a mixture of straw, clay and water, cow pats and maybe pebbles or other debris. The cob is compressed by foot to make structures that last hundreds of years. It is common in Devon and the southwest, though this example is in Nether Heyford in Northamptonshire (described by Christopher Wren in his 'TrogTrogBlog'). The heterogeneity of the material and irregularity of the surface creates many potential nest sites and its thermal properties, which make it a good, thick wall for cottages, make it also an ideal incubator for larval bees. This wall has a big 'aggregation' (large numbers of solitary bees nesting together) of *Anthophora plumipes*, along with *Melecta, Osmia bicornis* and *Andrena fulva.*

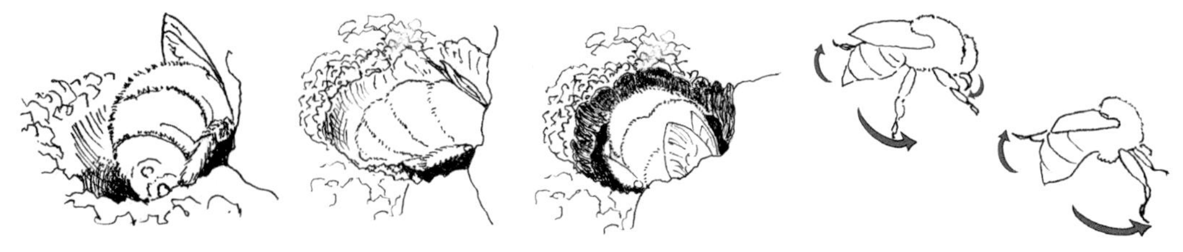

During hole construction the female uses rapid trembling movements of her tergites to tamp down with her pygidium soil particles that are pushed backwards with her fore and hind legs, whilst gripping with her mid-legs. At the same time, Dufour's gland secretions help to consolidate the tunnel walls. To clear the entrance area to the nest she uses an outward swinging motion of her hind legs. *Compare with* Dasypoda hirtipes*, p. 102.*

By contrast, *Anthophora pueblo* is making its nest holes in solid rock. Admittedly, a friable desert sandstone in Utah, USA, but very impressive nonetheless (newly described in 2016 by Michael Orr, et. al. in *Current Biology).* The bees seem to need water to help excavate the tunnels. The rock walls are honeycombed with tunnels which they hypothesise provide stable, re-usable nests, that, by being safe from collapse, also allow delayed emergence in extremely dry years. It is also possible that the rock makes it harder for parasites to exploit the nests. The bees avoid the hardest sandstones. In one site they use the walls of an ancient Pueblo cliff-dwelling site.

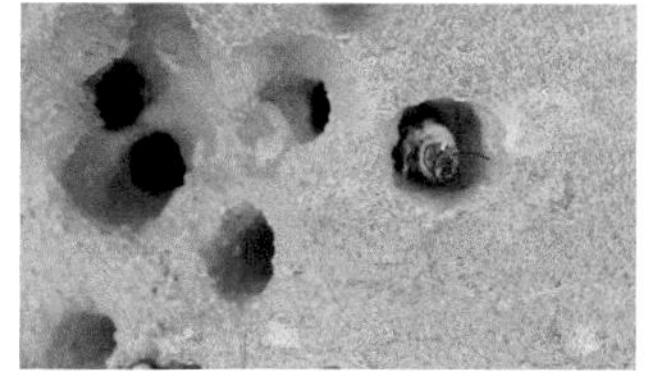

Our modern 'vegetarian' bees, which use pollen as a protein food, evolved from wasps. Wasps evolved as carnivorous animals, mainly using other insects and larvae as food for their young. Most modern wasps are still predators.

Fossils are not the only way to track evolution: **phylogenetics** aims to clarify origins using molecular DNA and morphological studies to create **cladograms** – diagrams that illustrate relatedness through geological time. A more detailed phylogeny for bees is shown on pp.141-142.

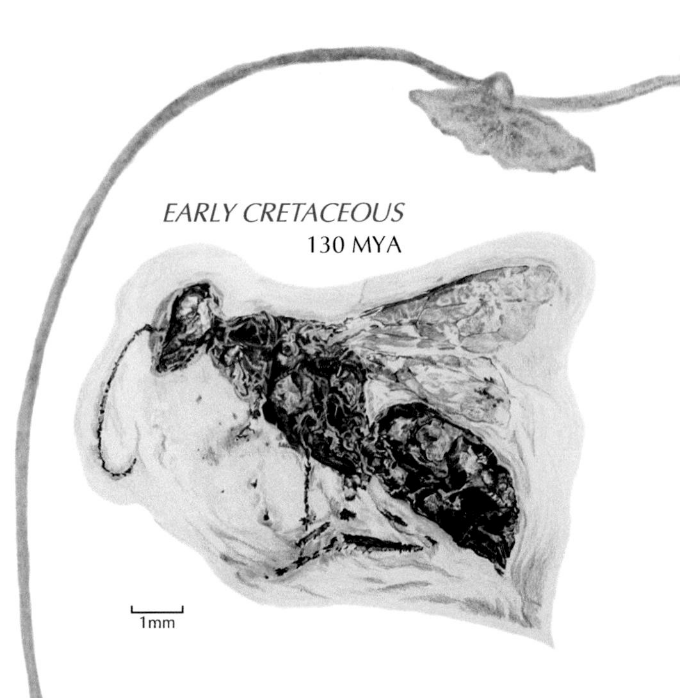

Before flowers evolved to use pollen to reproduce there were no true bees. Flowering plants (**angiosperms**) first evolved 130–145 million years ago (**MYA**) but their evolution accelerated from 90–115 MYA. This is known as the 'radiation' of flowering plants. The evolution of bees rapidly followed.

Phylogenetic research indicates that the first bees probably appeared around 125 MYA, though it could easily be as long ago as 145 million years. The fossil above is of a wasp from an extinct group that is close to the root of the modern bee families. *Angarosphex* lived from 130–145 MYA.

Fossil mandibles from the Early Devonian Rhynie chert from Scotland, possibly of the earliest known insect, *Rhyniognatha.*

Original drawings of fossil bees were made by the author from images in several sources: see "Notes on Illustrations: Source and References" at the back of the book.

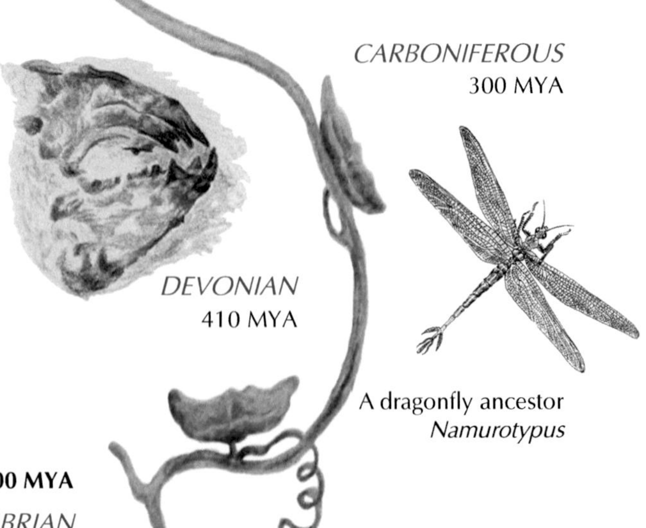

Microbial origins of life. **3,400 MYA**
ARCHAEAN/PRE-CAMBRIAN/CAMBRIAN

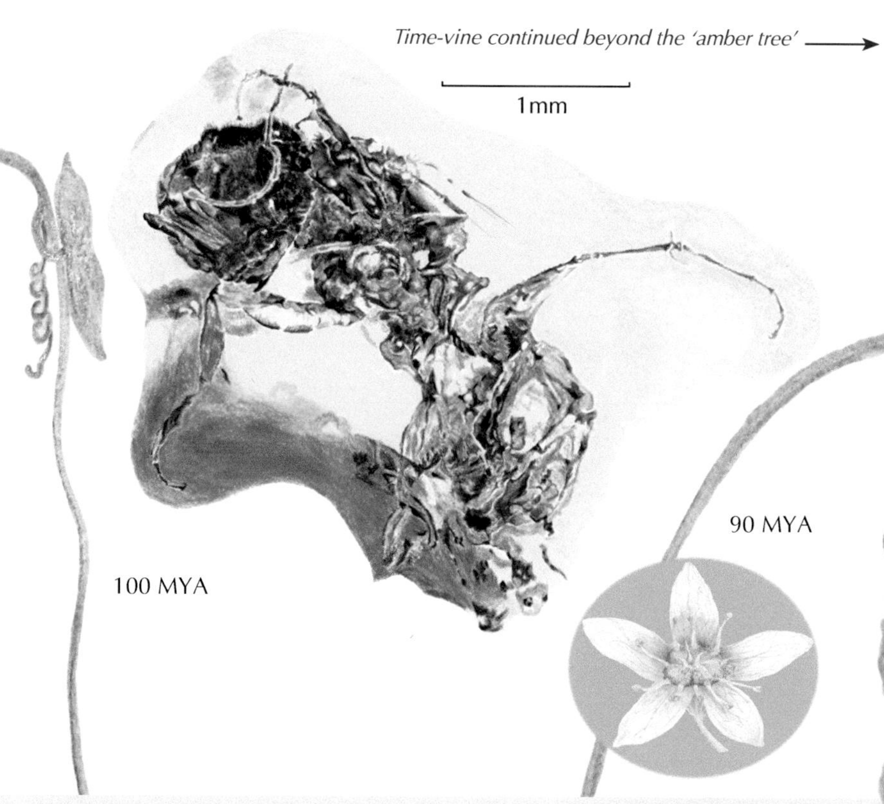

The fossil in amber above, *Melittosphex*, is from the middle Cretaceous period, about 100 MYA. Branched hairs on its head and leg show it to be close to bees, but it also has features that would make it a wasp. It was first thought to be the oldest bee species known but is now identified as a wasp precursor. The tiny 5mm flower is the fossil *Tropidogyne*, which was present as a species 96–110 MYA (a representation made with reference to a living relative, *Ceratopetalum*, from Australia). It was a possible source of forage for the first bees.

Amber. The best fossilised bees are found in amber. Sap seeping from trees attracts insects (*right*) and bees use it as nest material but also sometimes for sugars. They may become trapped in the sticky resin, which can then itself become buried and fossilised. Two of the fossilised bees shown on this and the next diagram are preserved in amber.

The bees survived the mass extinction when all the dinosaurs were lost at the end of the Cretaceous. *Cretotrigona,* the fossil bee in amber, *left*, was a social bee, like honey bees, though it was not a direct ancestor of our honey bee.

Palaeohabropoda, below, was a fossil ancestor of *Anthophora* bees. This fossil was found in volcanic deposits in the Puy de Dôme, France.

TERTIARY: PALAEOCENE

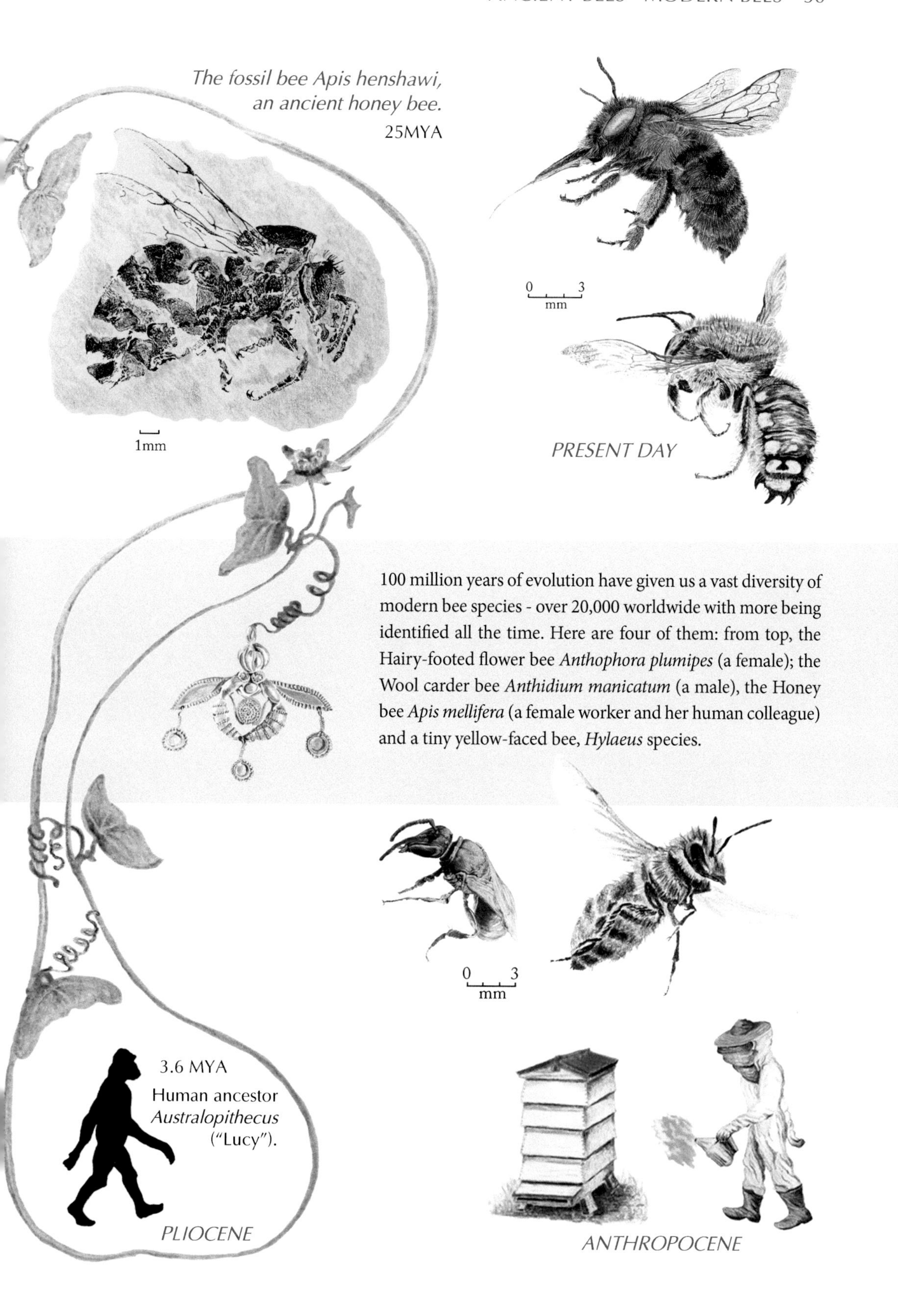

100 million years of evolution have given us a vast diversity of modern bee species - over 20,000 worldwide with more being identified all the time. Here are four of them: from top, the Hairy-footed flower bee *Anthophora plumipes* (a female); the Wool carder bee *Anthidium manicatum* (a male), the Honey bee *Apis mellifera* (a female worker and her human colleague) and a tiny yellow-faced bee, *Hylaeus* species.

All the Hymenoptera have a full metamorphic lifecycle called **holometabolism**, which means 'complete change'. Metamorphosis splits these insects' lifecycles into clear developmental phases, each with a distinct form and major change from one to the next. This enables the animal to exploit different habitats and food sources at each stage. The larvae of most holometabolous insects live in soil, in water or within or on plant tissue, and are very different to the adults that move freely to a variety of food sources and to find mates for reproduction. In some insects the adult stage is focussed entirely on reproduction and adults do not even feed. Other holometabolous insects are beetles, butterflies, moths, flies, caddisflies and lacewings.

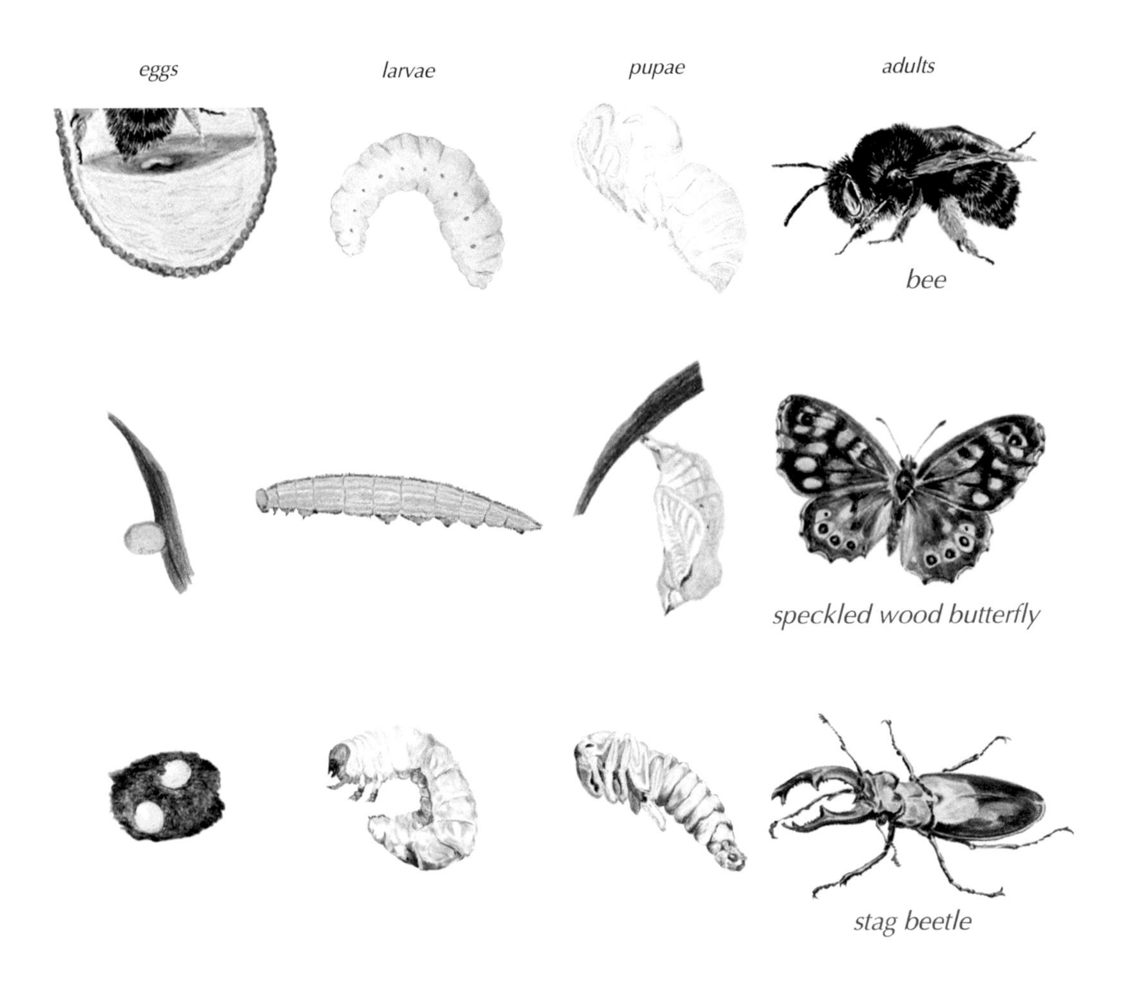

The larval body, built through continuous consumption, is transformed during the pupal phase of metamorphosis into the adult form. There is then no further growth (though some internal glands may still develop and mature). The larval and pupal stages allow development to be temporarily halted to withstand altered conditions. This is known as **diapause**, and may be in the form of **hibernation** over winter or **aestivation** (a type of torpor) over dry summers. A holometabolic lifestyle can therefore be very flexible and adaptable.

Not all insects are holometabolous – mayflies, dragonflies, damselflies, grasshoppers, crickets, earwigs, true bugs and some others are **hemimetabolous**: they do not have a larval stage like bees but instead progress smoothly from **nymphs**, their metamorphosis taking place more gradually. After a short-lived **pronymph** (a temporary grub-like stage) they quickly moult to resemble tiny adults. Then, through a series of further growth stages called **instars** (there are between four and ten instars in grasshopper and cricket species and between five and fourteen in dragonflies) and with a moult of the exoskeleton between each stage (**ecdysis**), the hemimetabolous insect gradually develops the adult form (the **imago**). Development in both holo- and hemimetabolous insects can also be brought forwards or slowed down in response to population pressure or predator or environmental stress, for example.

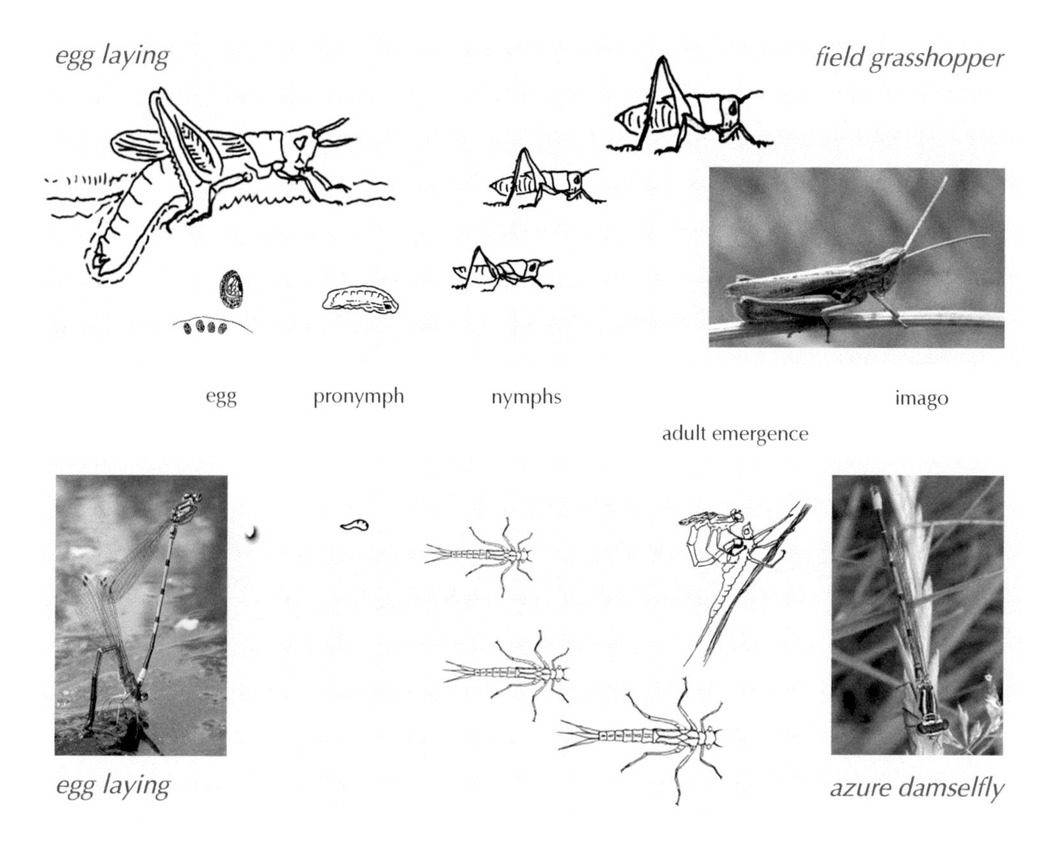

Holometabolism probably evolved about 300 MYA in the late Carboniferous period. The larval form may have developed from an extended pronymph stage of an early insect, a hypothesis supported by the pronymphs of hemimetabolous insects and the first instar larvae of the Holometabola: both have unhardened cuticles, both lack wing-buds and both have similarly reduced nervous systems and similar hormone levels.

The egg is laid onto a pollen/nectar mass in the nest cell and hatches quickly (**a**, below). In *Anthophora* the first stage of larval development takes place within the egg shell (the **chorion**). On hatching (**b**) the larva internally comprises mostly a digestive system. It also has a brain, nervous system and relatively simple respiratory system. It has simple mouthparts and antennae but no other external structures (**c**). Growth is through four instars after hatching, with corresponding moults. The larva consumes all the food left by the mother bee; *Anthophora* larvae then also eat the Dufour's gland secretion lining the cell.

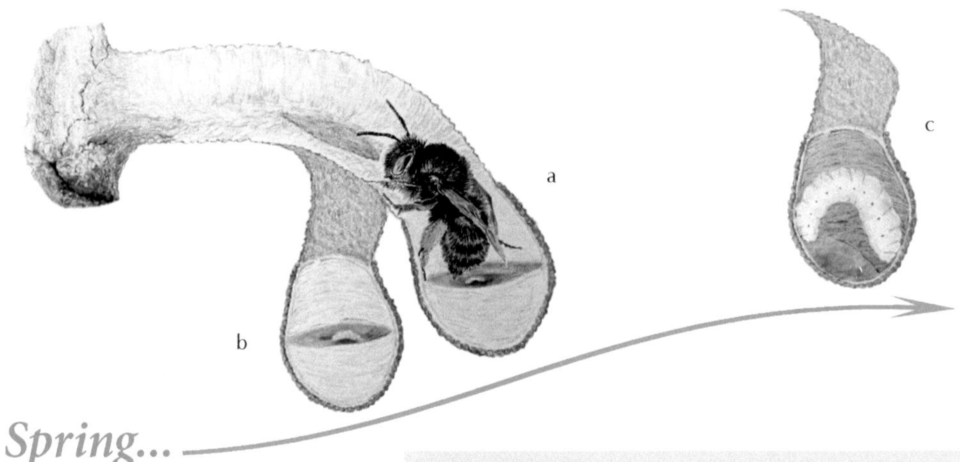

Imaginal discs:

The larva is the crucial stage of growth that enables later metamorphosis. This depends on specialised cells called **histoblasts**, which are situated in the larva's epidermis, just below the cuticle. Histoblasts are arranged in pockets called **imaginal discs**, which are unique to holometabolous insects. Imaginal discs correspond to mouthparts, antennae, eyes, legs, wings and genitalia. In the Hymenoptera the imaginal discs begin development in the embryo of the egg. During the metamorphic pupal stage of the Hymenoptera virtually all the adult cuticle is formed from the rapid growth and differentiation of these histoblasts. Larval structures are destroyed within the pupa (known as **apoptosis** or programmed cell death) and the fat bodies laid down during larval growth are used to fuel the rapid growth of the histoblasts. Metamorphosis into the adult form is controlled by the relative levels of two hormones, a **steroid** and the **juvenile hormone**. At the final moult, juvenile hormone levels are reduced dramatically, allowing the steroid to stimulate the shift to the pupal stage.

During growth the excretory tubules and midgut are closed off from the hindgut and store all waste, which maintains nest hygiene (**d**). Before the final moult the larva excretes the waste in one go before becoming a **prepupa** (**e**). The fifth instar, the prepupa, represents the beginning of the transformation to the adult form. However, a final moult to the pupal stage is required before all major adult features including wings begin to form. In the pupa (**f**) the cuticle also begins to darken until it reaches the adult state.

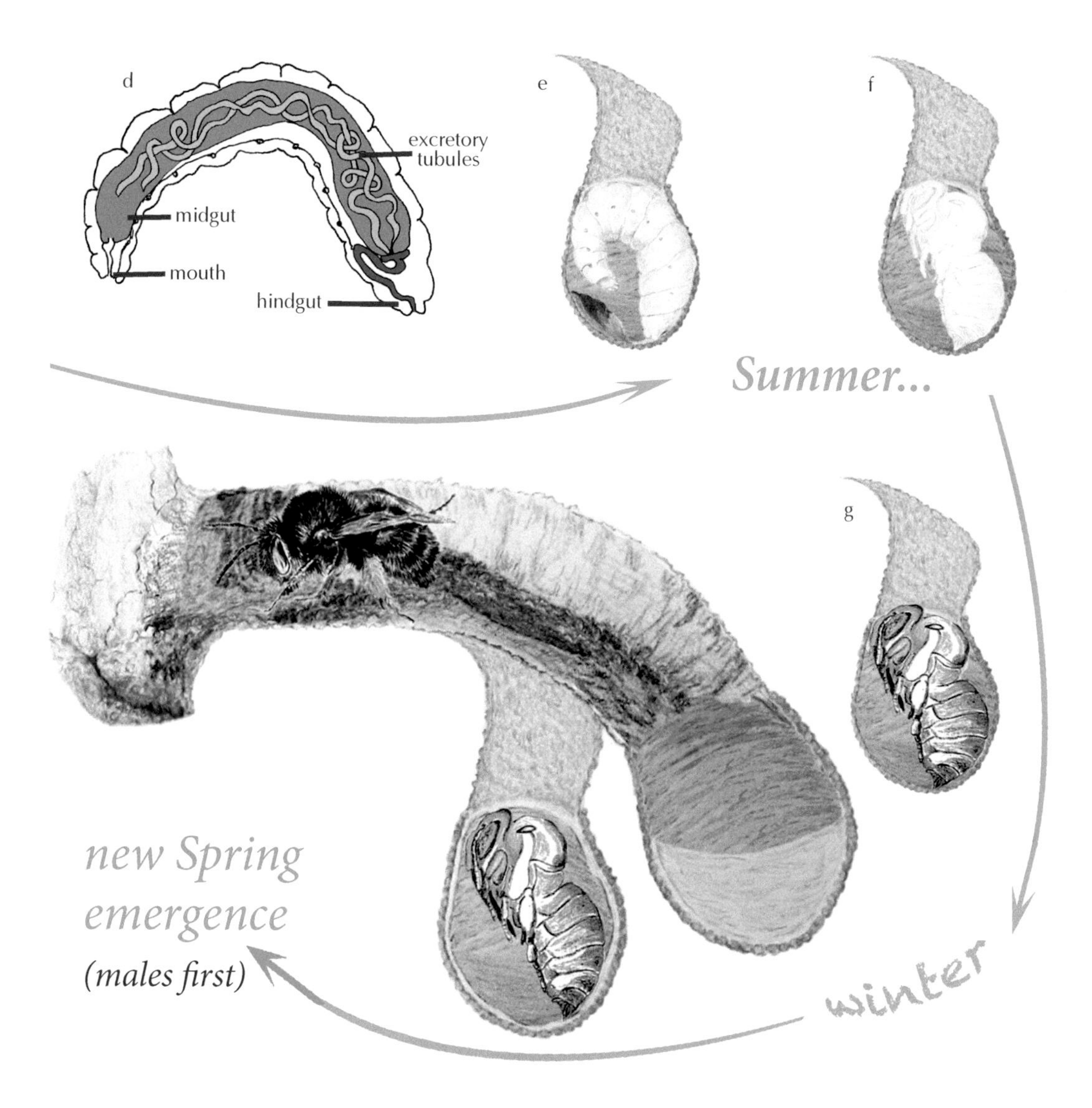

Anthophora plumipes overwinters in adult form (**g**) within a thin seal. Other bee species spin cocoons and overwinter as prepupae (mature larvae) or, more rarely, pupae. Under certain circumstances some species can extend the diapause for more than a year to overcome poor conditions.

Sex determination of the offspring of animals has evolved in varied ways. In mammals, the sex of offspring is generally determined by the presence of sex chromosomes from both parents. A male mammal can produce sperm after cell division with *either* X *or* Y chromosomes. The male himself has the XY combination to begin with; the male sex determination gene is on the Y chromosome. Since females have an XX combination of chromosomes, eggs have only X chromosomes, so when combined with male sperm the offspring can be *either* XX *or* XY. The XX offspring are females; the XY are males. A variant of this occurs in insects such as fruit flies. Sex determination may also be temperature dependent (as in alligators) or even due to changes in population structure (species of reef fish can change sex in response to an imbalance in the sex ratio of a social group).

However, in the Hymenoptera, males hatch from an unfertilised egg and so only have chromosomes from the mother bee. These males are called '**haploid**' – they have no father and so are genetically identical to the mother. Females hatch from fertilised eggs and have chromosomes from both parents – they are '**diploid**'. In honeybees, fertilisation can be with sperm from many males, though most solitary bees mate only once. The determination of sex is thus dependent on the presence or absence of chromosomes from each of the parents. This system is called '**haplodiploidy**'.

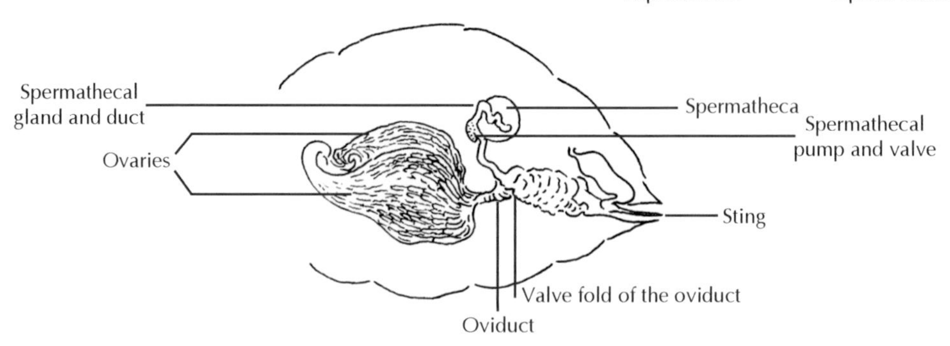

Haplodiploidy is made possible by the ability of females to store live sperm in an internal sac, a characteristic of many insects. After mating, sperm travels via the **spermathecal duct** to the **spermatheca** (above). Secretions from the spermathecal gland keep sperm alive for the rest of the female's life, which can be for several years in honeybees and decades in some ants. In one sense, the female is hosting cells from another animal – a form of **endosymbiosis**. For most solitary bees this will be a shorter period, though some species will mate before overwintering. After an egg has moved from the ovary down the oviduct it reaches the valve fold, where the female can selectively release or withhold sperm via the **spermathecal pump** and **valve.**

Following the evolution of haplodiploid sex determination and female control of fertilisation, complex behaviours and adaptations have developed. The highly social honeybees (p.60) have a female-only worker generation with males produced only for colonial reproduction around swarming time. In solitary bees, haplodiploidy enables the sex allocation of eggs to respond to environmental pressures such as the presence of parasites that enter open cells (the fly *Cacoxenus indagator,* below right, for example), poor weather or foraging conditions and the progress of the season with ageing and deterioration in the building female.

The risk of a fly parasitising a cell during construction is greater for cells nearer the entrance: the fly can easily enter and lay whilst the female is away. To avoid this, the female bee spends less time foraging for these cells, deposits less food in them and then lays a male egg inside them (males are smaller than females and male larvae need less food). More food is provided to the more secure cells at the back of the nest, where female eggs are laid. The body sizes of offspring hatched near the back of the nest are generally larger than those further forward, because the mother bee gradually reduces her foraging time as she completes the nest hole. For solitary bees using existing holes (such as mason and leaf-cutter bees), smaller hole sizes may produce single sex male nests. This should be borne in mind when designing 'bee hotels': too many holes of the same size could skew the local sex ratio of the bees. Females often make several nests using holes of varied entrance sizes. *Osmia bicornis,* for example, often nests in markedly different hole sizes (below). On the right is a waiting *Cacoxenus indagator.*

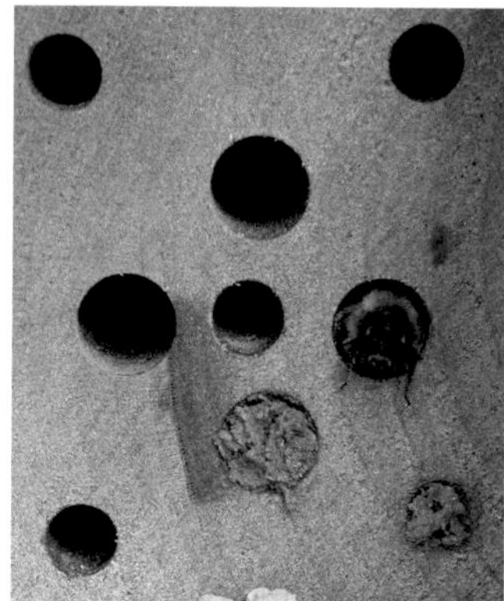

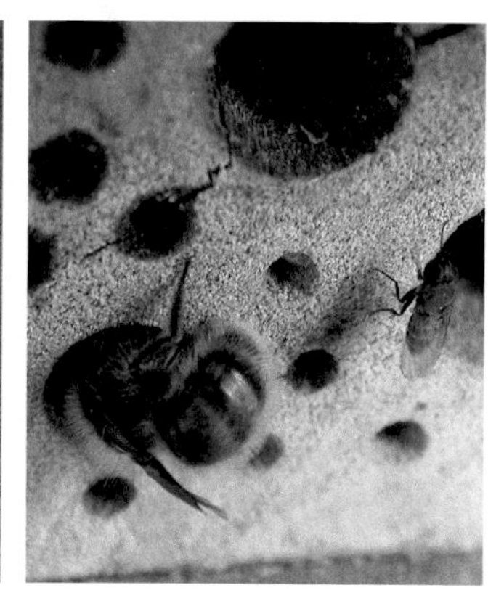

Evolution of haplodiploidy

Haplodiploidy also determines sex in the thrips (Thysanoptera) and in some species of true bug (Homoptera), beetle (Coleoptera), mites and ticks (Acarina). There may be 20 separate evolutionary origins for haplodiploidy. Multiple evolutionary origins suggest a variety of pathways that could lead to it and there is much debate as to how it actually happened in the Hymenoptera. It may have originated from endosymbionts – gut bacteria in the rotting wood-consuming larvae of ancestral sawflies that controlled inheritance of the male genome. Haplodiploidy could then persist through a complex net of ecological advantages.

'singular sons - doubled daughters'

When we describe a person as solitary we usually mean that they live a life with little company; we describe people as 'lonely' if they would prefer not to do this. A hermit is someone choosing to live alone – perhaps to meditate, write or to study philosophy, religion or nature – as Henry David Thoreau did at Walden Pond, (above, centre). 'I am no more lonely than a single mullein or dandelion in a pasture, or a bean leaf, or a sorrel, or a horse-fly, or a humble-bee,' he wrote.

Apart from the odd political ideologue most humans would recognise that there is such a thing as society. Most of us also value the social networks that constitute societies. We organise ourselves into communities – often with distinct boundaries – and form social networks within and beyond those communities for particular purposes, for example by joining the Royal Entomological Society, the RSPB or the International Society for Endangered Cats. People often like to live in groups of different sizes including in large buildings such as the Trellick Tower flats in north London (above, left). The wooden Trellick Bee Tower alongside the flats and in its wildlife garden home (above right) is a replica of the block. It has holes drilled for solitary bee nests (see pp.77–78 and 135–136) and scaffolding planks that represent the balconies of the flats. But many humans also prefer much smaller groups and more intimate living conditions – they find big-city life (a very recent phenomenon, after all) overwhelming.

It is easy to see that 'society' could never be a simple, easily defined concept or object. Is there any connection between the behaviour of African hunting dogs, the argumentative sparrows in the back garden, a pod of killer whales or the members of the Bees, Wasps & Ants Recording Society (BWARS)? Long-term studies of animal social systems are beginning to demonstrate the diversity of organisation but also the similarities and commonalities to be found across species, even groups, of organism. We can learn something about human societal structures from our studies of cetacean behaviour, for example, despite the evolutionary distance between us. One wonders how much farther social insects are from human societies.

We call bees solitary to distinguish them from social bees nesting in large colonies with a queen and hundreds, perhaps thousands, of cooperating female worker bees. It is, perhaps,

Between 50 and 60 nest holes in an aggregation of the ashy mining bee Andrena cineraria, *in a sandy bank in Northern Ireland.*

a slightly misleading term, since the Hymenoptera, a vast group of insects, are bound to have a wide range of life histories after over 100 million years of evolution. Large numbers of solitary bees nesting together are often called an aggregation, for example the ashy mining bees above. They are a 'community of association' and may or may not be cooperating in any complex way, according to species. In aggregations of *Anthophora* and *Andrena*, females are seen entering existing nest holes that may already be occupied or abandoned. Nest holes are often reused in this way – and sometimes females may usurp other females from nest holes.

Living close to each other has many advantages: male bees can very quickly find females to mate with and females can save time looking for holes if they use popular sites. They could also reduce their individual chance of being parasitised by flies and wasps (safety in numbers) and can perhaps more effectively defend their nests against such parasites and predators. On the other hand, a large mass of breeding insects is an attractive food source for parasitic species, which can themselves become very abundant, and for predators (hornets, badgers, honey buzzards). Mating and breeding success may be the driver for much social evolution be it in whales, humans or insects, but the consequences of living together in complex social colonies are many sophisticated adaptations.

Highly evolved social systems in animals focus on the care of young, especially when offspring are slow to reach maturity and require continual provisioning and defence against predators (sperm whales and humans are example species). Bees put far more effort into rearing young than, say, butterflies. They construct complex nests and lay relatively few eggs in them. Purely solitary bees then rely on this investment in the nest to keep the eggs and larvae safe and fed – the nest cell is filled with food (mass provisioning) and the young are left to grow alone. In contrast, honeybees and ants, which build even more complex nests, use non-breeding workers to continually feed the growing larvae and store food for security. Colonies can survive for many years.

Most bee species (90%) are solitary – they nest singly or in loose aggregations across suitable habitat (e.g. *Andrena bicolor*, top left). There are many examples, however, of species with life cycles that approach the highly social honeybees. Bees that habitually nest in large aggregations are solitary but **communal** (e.g. *Andrena cineraria*, p.58), perhaps with hole-sharing taking place but without cooperative behaviour between females of the same generation. Then there are species with females that stick around to defend (e.g. large carpenter bees such as *Xylocopa* species) and those that sometimes feed the young continuously through development (some with assistance from sisters or non-breeding female young, such as small carpenter bees, *Ceratina* species). The latter are called '**sub-social**' (or 'cooperative'). Some researchers have also recognised categories between communal and sub-social, such as 'quasi-' and 'semi-social' systems (perhaps in tropical/sub-tropical environments). **Primitively eusocial** bees are those whose lifecycles begin with a solitary female founding the nest, and progress with the season to producing non-breeding females that first share and then take over the role of foraging and feeding larvae. Examples would be bumblebees and some members of the Halictidae (e.g. *Lasioglossum morio*, above right, a male). The continuum from solitary to social species is illustrated by species that seem ambiguous, such as *Halictus rubicundus*, above left, which is solitary in the north and social in the south (see pp.61-62).

Eusocial means 'perfectly social' and is applied only to animals that share the care of their young, that cooperate across generations, that have a sterile, non-breeding worker class and that have one or sometimes several queens responsible for producing young. They often maintain large colonies. The most sophisticated (**advanced eusocial**) species are ants (Formicidae, over 14,000 species), termites (over 2,000 species), honeybees (*Apis* – currently seven species are recognised), the stingless bees of the tropics (Meliponinae, around 1,000 species) and some wasps (over 80 species of Vespidae and one species of Sphecidae). In insects, eusociality has evolved many times – around ten times in Hymenoptera species with a sting (aculeate), six to eight times in bees but only once in the Apidae. That eusociality has emerged this many times in the aculeate Hymenoptera suggests the existence of predispositional factors, which could be complex nest construction, continuous provisioning of the nest or reuse of nests by later generations leading to sub-social behaviour. Haplodiploidy is no longer considered an important factor – all members of the Hymenoptera are haplodiploid but aculeates comprise only 46% and the majority are still solitary.

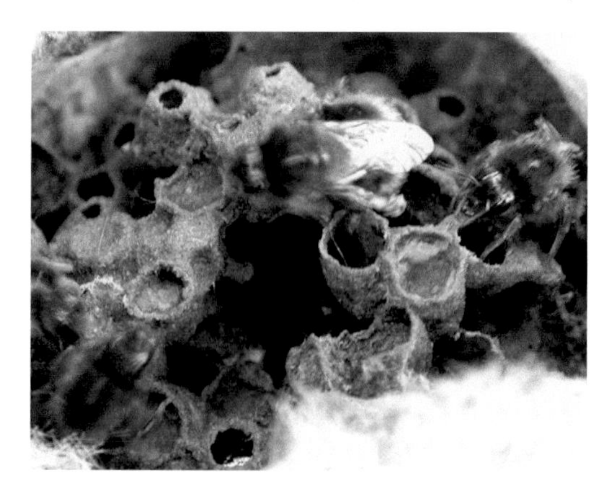

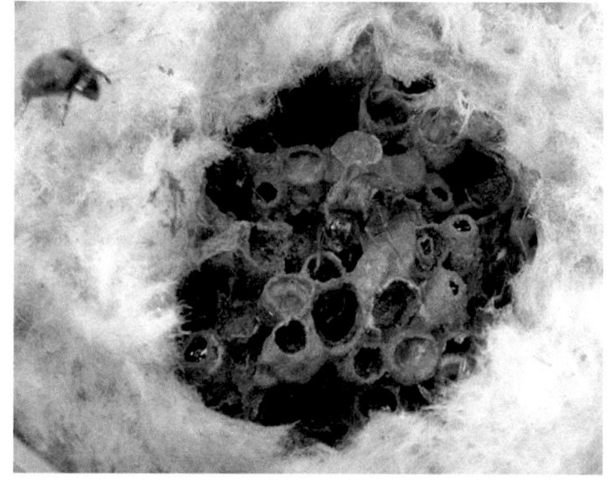

The primitively eusocial bumblebee lifecycle (*Bombus hypnorum* nest above) is solitary at the start of the season and eusocial at the end. The queen alone forages for, cares for and feeds the first generation of female larvae. These become workers and the nest rapidly grows as they feed successive generations. There are no males in nests (and thus no mating) until later in the season when new queens are also reared.

The classic eusocial hymenopteran nest is that of the honeybee *Apis mellifera* (above right, queen marked with blue spot). Honeybees have evolved complex environmental monitoring and communication systems that facilitate food storage and preservation, the environmental control of nests and sophisticated colonial reproduction using swarming and consensual nest-site choice. All ant species are highly eusocial and some have distinct worker sub-castes: minor workers, majors ('soldiers') and even super-majors.

That life is unfixed, mutable, constantly adjusting or shifting from one state to another was appreciated in the sixth century BCE. Heraclitus of Ephesus saw life as being in eternal flux, with reality constantly replacing itself in a series of transformations. Darwin and Wallace, sometime later, approached life with similar thoughts and framed what has now become our understanding of the theory of evolution by natural selection. Bees, complex as they are, cannot be categorised simply as solitary, communal, primitively eusocial or highly eusocial. Time and space give them opportunities to be anything but fixed.

*Members of the family Halictidae are called 'furrow bees' after the groove ('**rima**') in tergite 5 - see also p.90.*

The attractive orange-legged furrow bee *Halictus rubicundus*, above, is of great interest to evolutionary biologists because it has both solitary and eusocial lifecycles. In the northern parts of its range it is solitary and often nests in aggregations. This female was part of a very small aggregation in a garden by the Ballinderry River in Northern Ireland. Here, a female, having mated the previous autumn and overwintered, emerges to become a '**foundress**', digs a new burrow and lays a first brood in late April and May (**A**, opposite). Females from this brood will mate in the summer and become the following year's foundresses. In southern Britain, however, some of the first-brood females do not mate, become workers instead and rear a second brood. This brood is laid by the foundress female or, if she dies, by a mated first-brood female: the nest is primitively eusocial (**B**, opposite). This *Halictus* has a wide range across the northern hemisphere and a lot of research into the '**plasticity**' of its lifecycle has been done in North America as well as in Europe. (Plasticity here means the ability within a species to shift from social to solitary and back, according to environmental conditions.) In the US, earlier work suggested a genetic component to this difference – possible speciation – or evolution in action. More recently, experiments in Britain and Ireland moved mated females from northern sites (Belfast/Penrith/Peebles) to southern sites (Wicklow/Sherborne/Sussex/Cambridge) and vice versa. Females moved north from established eusocial sites to solitary aggregations changed their behaviour and established solitary nests. The reverse happened with the bees moved south. Further investigation has concluded that other variables such as body size, the demography of populations and foraging strategies are all plastic in this species and that there is little evidence here of genetic differences between social and non-social populations. Differences in these variables would be expected: social nests with two generations require a longer season and foundress females to be active earlier than would be possible in the north. Further, foraging patterns would be different when many worker females are involved rather than one foundress female. The species is responding to environmental conditions. This sociality is often called '**facultative**' (as opposed to the '**obligate**' sociality of honeybees, whose behaviour remains eusocial in an extremely wide variety of environments).

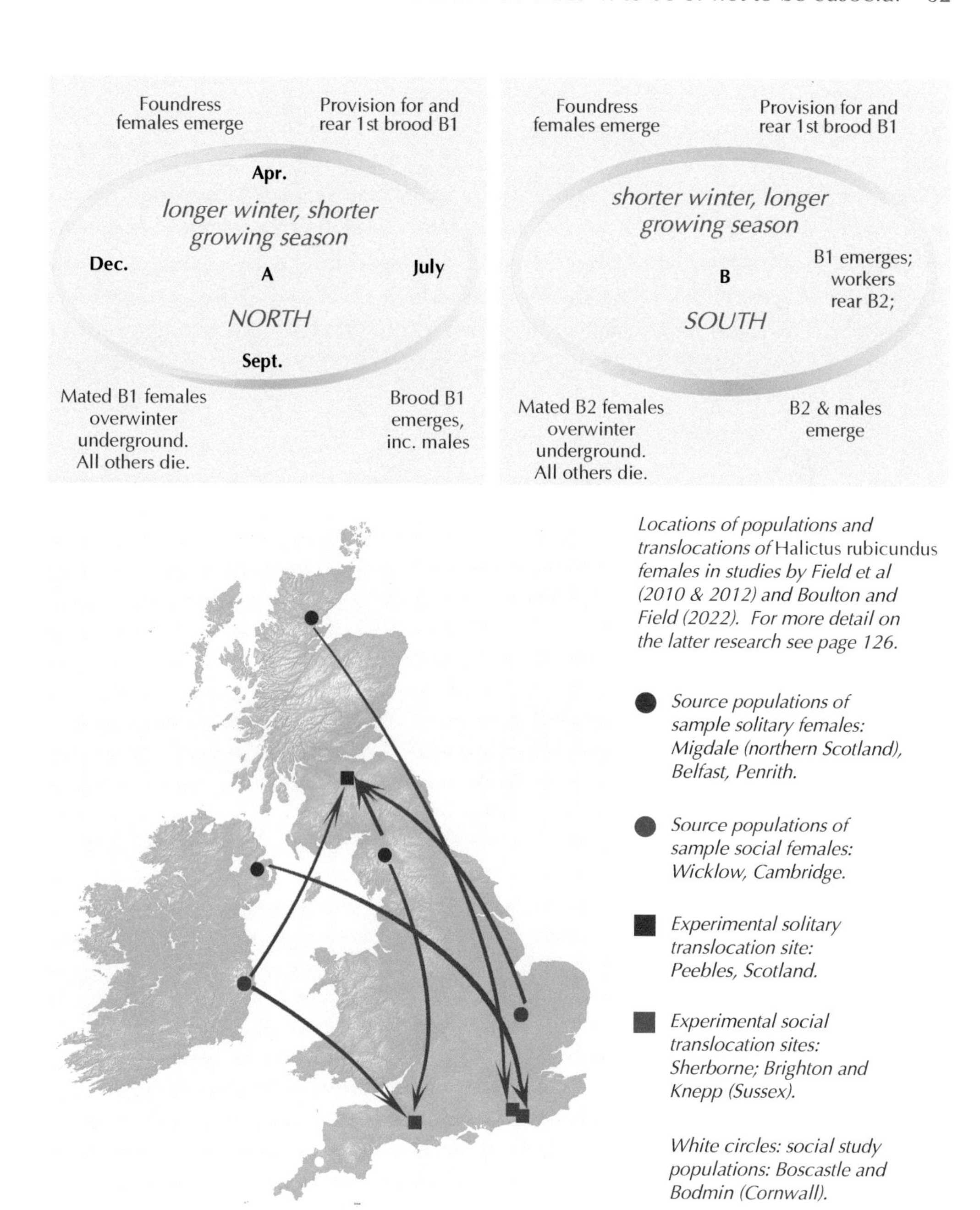

Locations of populations and translocations of Halictus rubicundus *females in studies by Field et al (2010 & 2012) and Boulton and Field (2022). For more detail on the latter research see page 126.*

This plasticity in social behaviour could suggest a relatively recent origin for the behaviour. Studies suggest that Halictidae eusocial evolution took place three times between 20 and 22 MYA (interestingly, in periods of global warming). However, reversion to sub-social or solitary lifecycles has occurred up to 12 times, with some species currently having nests that may be both eusocial and sub-social (i.e. **polymorphic** – social and solitary in the same populations). The more stable eusocial honeybees evolved much earlier – around 65 MYA. See pp.125-126 for more on social plasticity and climate change.

Another very large genus in the Halictidae, *Lasioglossum* (see pp.93-94), also contains primitively eusocial species. A recent phylogenetic comparison of *Halictus* and *Lasioglossum* suggests a single origin for eusociality in these two genera, pushing it back to around 35 MYA: as our knowledge increases it is likely that interpretations may change again. Research on a common species, *Lasioglossum calceatum*, also using transplantation experiments moving females from northern to southern populations in Britain, has shown very little plasticity in the species. The two populations appear to be genetically distinct. *Lasioglossum calceatum* (below left, a male foraging on golden rod in London) seems to have lost its social polymorphism in the UK, whereas *Halictus rubicundus* retains its plasticity. In North America, *Halictus rubicundus* shows genetic differences across its range, as with *Lasioglossum calceatum* in the UK. Clearly there is great scope for future research.

Another primitively eusocial species, Lasioglossum morio*: a female foraging on a fleabane flower.*

The evolution of advanced eusocial species in the Hymenoptera is such that their colonies have been called 'superorganisms' by Edward Wilson and others: the queen can be regarded as the reproductive organ – the '**germ cell**' – and the workers would then be '**somatic cells**'. Workers are 'disposable soma' and have much shorter lifespans than the queen. In a superorganism such as a honeybee or ant colony the workers could also be seen as mobile 'sensory cells' that can be sent widely across the environment to bring back resources and information to the colony. As with a human city, the hymenopteran colony is able to efficiently exploit the surrounding environment to promote its own growth. This does not lead to over-exploitation (unlike many human cities) – the ant colony will move its base; the bee colonies are widely spread. Too many bee hives kept in one apiary, though, can have consequences for the ecology of the surrounding area.

There are also primitively eusocial thrips and aphids, snapping shrimps, a beetle species, a mammal (the naked mole rat) and possibly one spider. No mammal, though, even the human, is a highly eusocial superorganism like ants or honeybees. Human sociality evolves in other unique, perhaps damaging, ways.

Germ cells: reproductive cells in an organism, perpetuated during reproduction. The honeybee queen is the longest living member of the colony. Workers rarely live beyond one season; queens can survive for three years or more. During 'colonial reproduction' (swarming) the resident queen departs with part of the 'somatic' workforce to establish a new colony.

Somatic cells: all the cells in the body except the reproductive cells. Worker honeybees fulfil many of the roles for the colony that somatic cells (including in the form of organs) would fulfil in an organism, such as food transport and processing, temperature control, sensory data collection and communication, defensive systems. Workers have a much shorter lifespan: they are constantly lost and replaced, as with somatic cells in an organism.

The notion of the colony as a superorganism with queens as 'germ plasm' and workers as 'soma' was introduced by Willliam Morton Wheeler in 1911. The history of ideas, part of that Heraclitean eternal flux, and the massive research effort put into honeybees and other social insects, has led to many interpretations and reinterpretations of these notions. Tom Seeley, for example, discusses the 'honey bee colony as a unit of function' in his introduction to *The Wisdom of the Hive: the Social Physiology of Honey Bee Colonies* (1995). He followed this with his study of the honeybee swarm and how the honeybee finds a new nest site, including a discussion of 'The Swarm as Cognitive Entity' in *Honeybee Democracy* (2010), referencing Hofstadter's late-1970's book *Gödel, Escher, Bach: An eternal Golden Braid.* Hofstadter considered ant colonies as being little different from brains. Evidently, there will be no shortage of future reinterpretations, given that observing, investigating and trying to understand the complexity of social insect colonies has been a human focus for over 2,000 years.

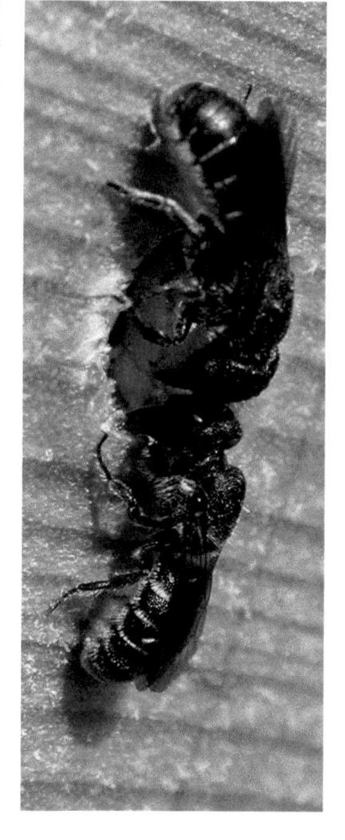

Two female resin bees Heriades truncorum *at work sealing the same nest without conflict. Confusion or cooperation? See p.90 for more on this species.*

Consideration of the demands of sociality (individual recognition, sophisticated communication requiring good sensory information and memory) has produced the 'social brain hypothesis', whereby in primates and other mammals brain development (size and complexity) is linked to social behaviour. Insects are excellent organisms for testing the hypothesis as they can have complex social behaviour with relatively simple nervous systems. For example, the brains of breeding facultatively eusocial female tropical sweat bees (*Megalopta genalis*) are much larger in cooperatively breeding females than in workers or females breeding solitarily. Honeybees, however, with their complex 'waggle-dance' behaviour and with the colony usurping some of the roles of the individual brain, do not have specific brain adaptations compared with other non-dancing eusocial bees. Nor do social wasps.

The OVIPOSITOR - NOT A STING

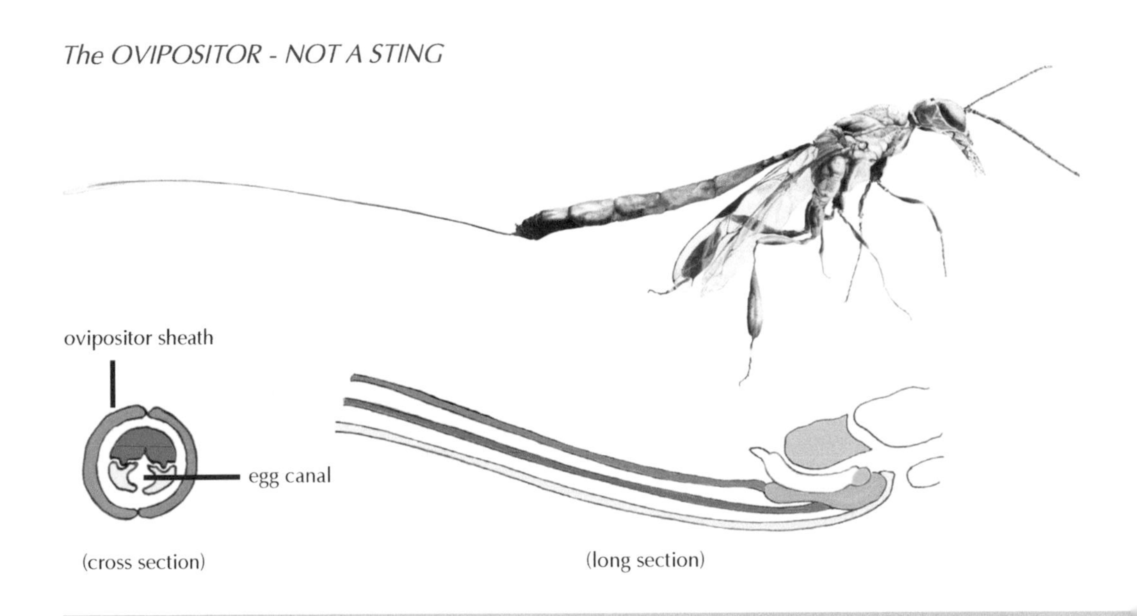

The PREDATORY STING

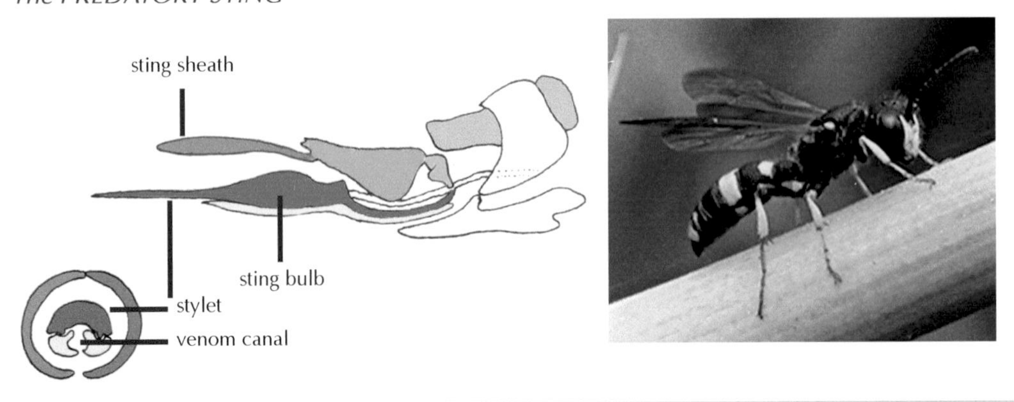

The DEFENSIVE STING

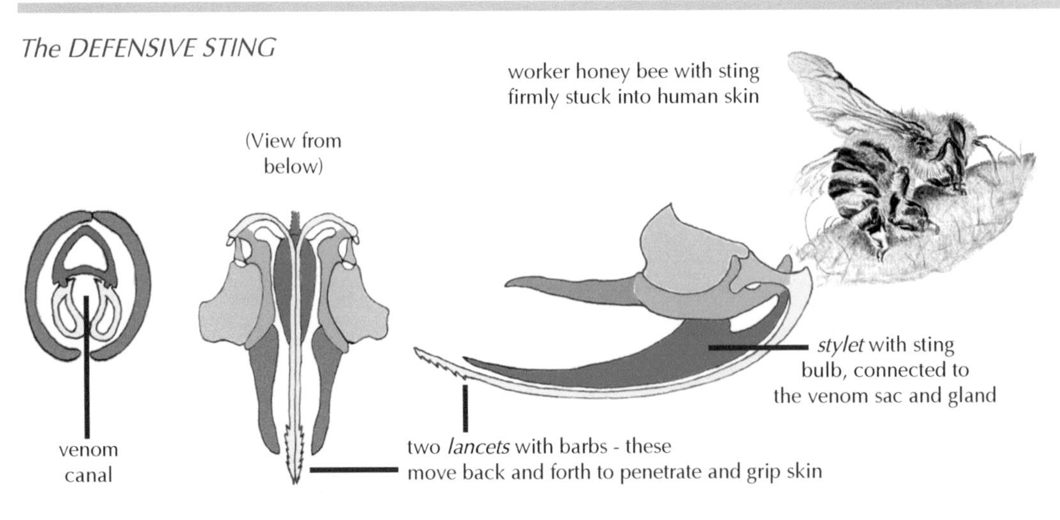

Ants, bees and wasps evolved a narrow constriction – the petiole – between the propodeum and the rest of the abdomen (p.9, the 'wasp waist'), from early Jurassic times, c.195 MYA.

The **ovipositor** (egg placer) and petiole give flexibility and enable a female wasp to find and penetrate a living host insect (perhaps hidden within leaves or wood) for its egg and young. This **parasitoid** lifestyle evolved from around 190 MYA. True parasites do not kill the host during growth, whereas a parasitoid lifestyle more closely resembles predation – the host is ultimately killed. Wasps very quickly evolved parasitoid lifestyles. The wasp on the left is *Gasteruption jaculator*, which lays its eggs in solitary bee nests (see p.103).

Above, a female parasitoid ichneumon Ephialtes manifestator *at the Trellick Bee Tower.*

The **sting** evolved from a modification of the ovipositor. Eggs are now laid from an opening beneath the sting. A female paralyses or kills her prey before feeding it to the young. Many solitary wasps paralyse prey and store it in a nest or burrow. The still-live prey remains a fresh food supply until the egg hatches, and as the larva feeds it kills the prey. The wasp *Cerceris rybyensis* (left) hunts for solitary bees to stock its nest. Other wasp species use honeybees, flies, bugs and spiders as prey (see pp.105-106).

Above, a female spider-hunting wasp Anoplius nigerrimus *predates a fearsome louse-eating spider (*Dysdera*). She has paralysed the spider and will carry it to her nest as food for her larvae.*

With the evolution of large, social colonies came the evolution of the sting as an effective defence. Social wasps feeding their young on other insects and their larvae use the sting to capture prey as well as to deter nest predators. As a hunting tool, the sting is a relatively simple spear. For the 'vegetarian' highly social honeybee *Apis mellifera* the sting has become a complex predator deterrent. The barbs on the lancets of the sting grip mammalian skin and work to and fro to push deeper into the flesh, ensuring more venom is inserted. Eventually, the struggle damages the bee's abdomen, leading to death. This damage releases more alarm pheromones (scents), attracting more bees to the attack until the intruder is driven off (see next page). The venom gland remains attached and continues to pump venom into the wound. The pheromone signal then rapidly disperses: workers have been lost but the colony has been defended.

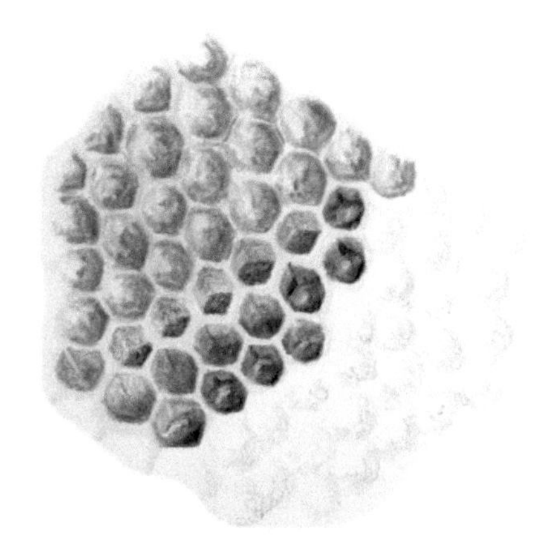

Honeybee: builds a large, long-lasting nest in a hollow tree or bee hive. There are thousands of larvae plus honey and pollen food stores in honeycomb, which is made of wax. These are good food sources for mammals (including bears, badgers, humans and mice in the winter), birds (woodpeckers and, for honeybee species nesting in the open, honey buzzards), wasps and hornets. The predation threat is ongoing.

Social wasp: builds a large nest in a hole in the ground, in a tree, in a shed or roof space. The nest is often made of wood pulp processed into paper. It dies in the autumn. It contains thousands of larvae but no food stores. The larvae are fed progressively on masticated insect food and are themselves good food sources for mammals, birds and other insects in summer.

Solitary bee: each nest contains only a small number of larvae, which are protected in the substrate by careful nest construction. Female bees die after building their nests (living for only four to six weeks) and males die not long after mating. Birds such as tits or woodpeckers may attack nests, perhaps in the winter, but during construction by the females there is little need for defence. The females have stings but only to use against insect predators, parasites or spiders. They have not evolved defensive or aggressive behaviour towards large animals and will not generally sting humans.

Two views of Anthophora*'s sting. She has no barbed lancets but will give you a sharp, painful stab should you mishandle her.*

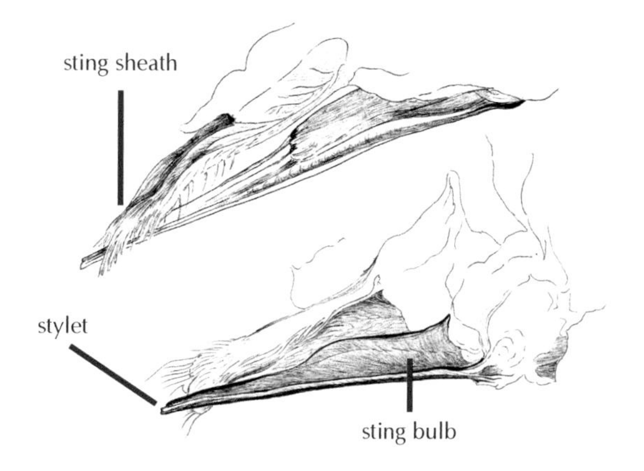

The sting has evolved from the ovipositor and so **male bees** have no sting. In species with territorial or highly competitive mating strategies, however, aggressive behaviour and other weapons are found, notably in the male wool carder bee (p.79–80).

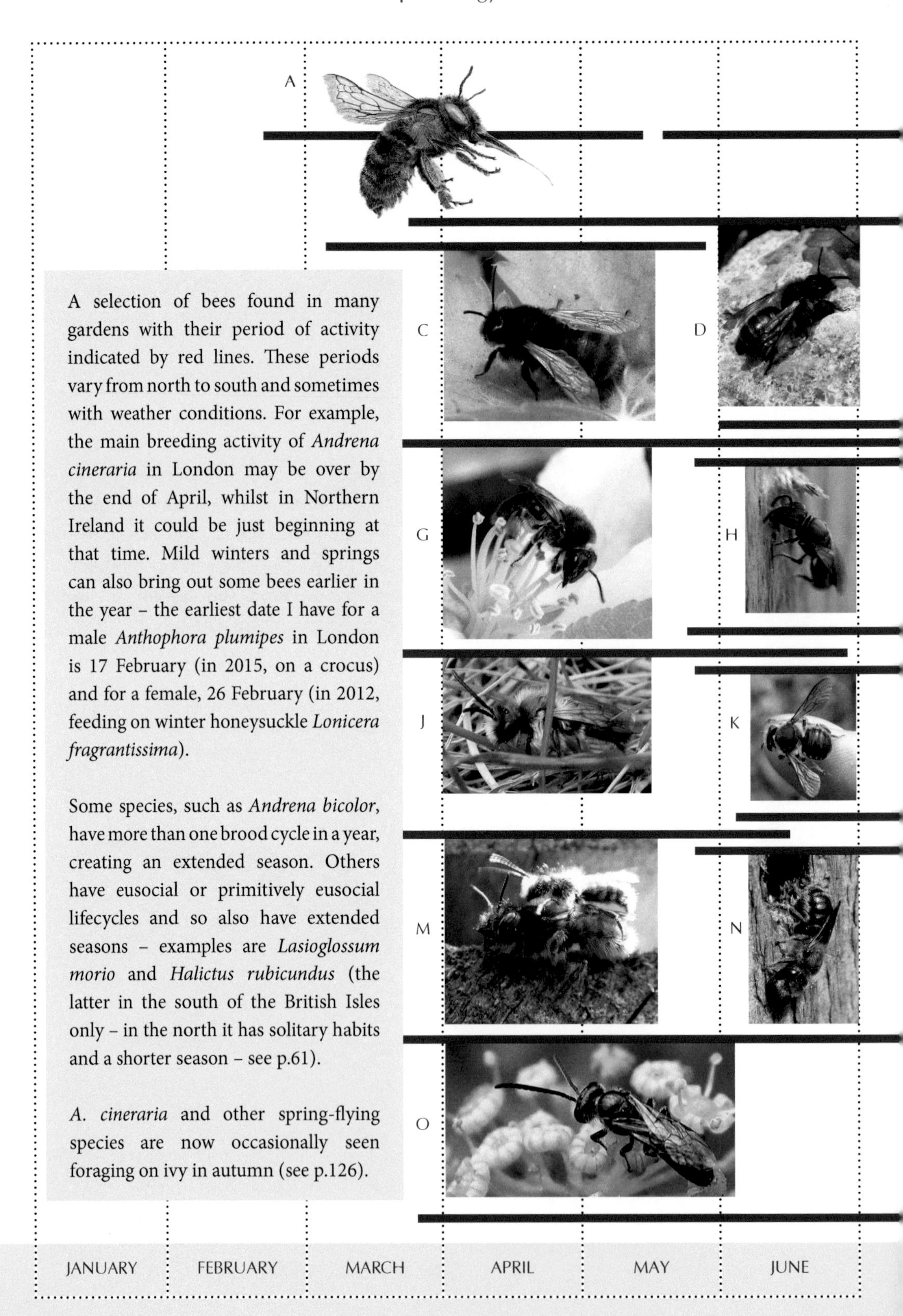

A selection of bees found in many gardens with their period of activity indicated by red lines. These periods vary from north to south and sometimes with weather conditions. For example, the main breeding activity of *Andrena cineraria* in London may be over by the end of April, whilst in Northern Ireland it could be just beginning at that time. Mild winters and springs can also bring out some bees earlier in the year – the earliest date I have for a male *Anthophora plumipes* in London is 17 February (in 2015, on a crocus) and for a female, 26 February (in 2012, feeding on winter honeysuckle *Lonicera fragrantissima*).

Some species, such as *Andrena bicolor*, have more than one brood cycle in a year, creating an extended season. Others have eusocial or primitively eusocial lifecycles and so also have extended seasons – examples are *Lasioglossum morio* and *Halictus rubicundus* (the latter in the south of the British Isles only – in the north it has solitary habits and a shorter season – see p.61).

A. cineraria and other spring-flying species are now occasionally seen foraging on ivy in autumn (see p.126).

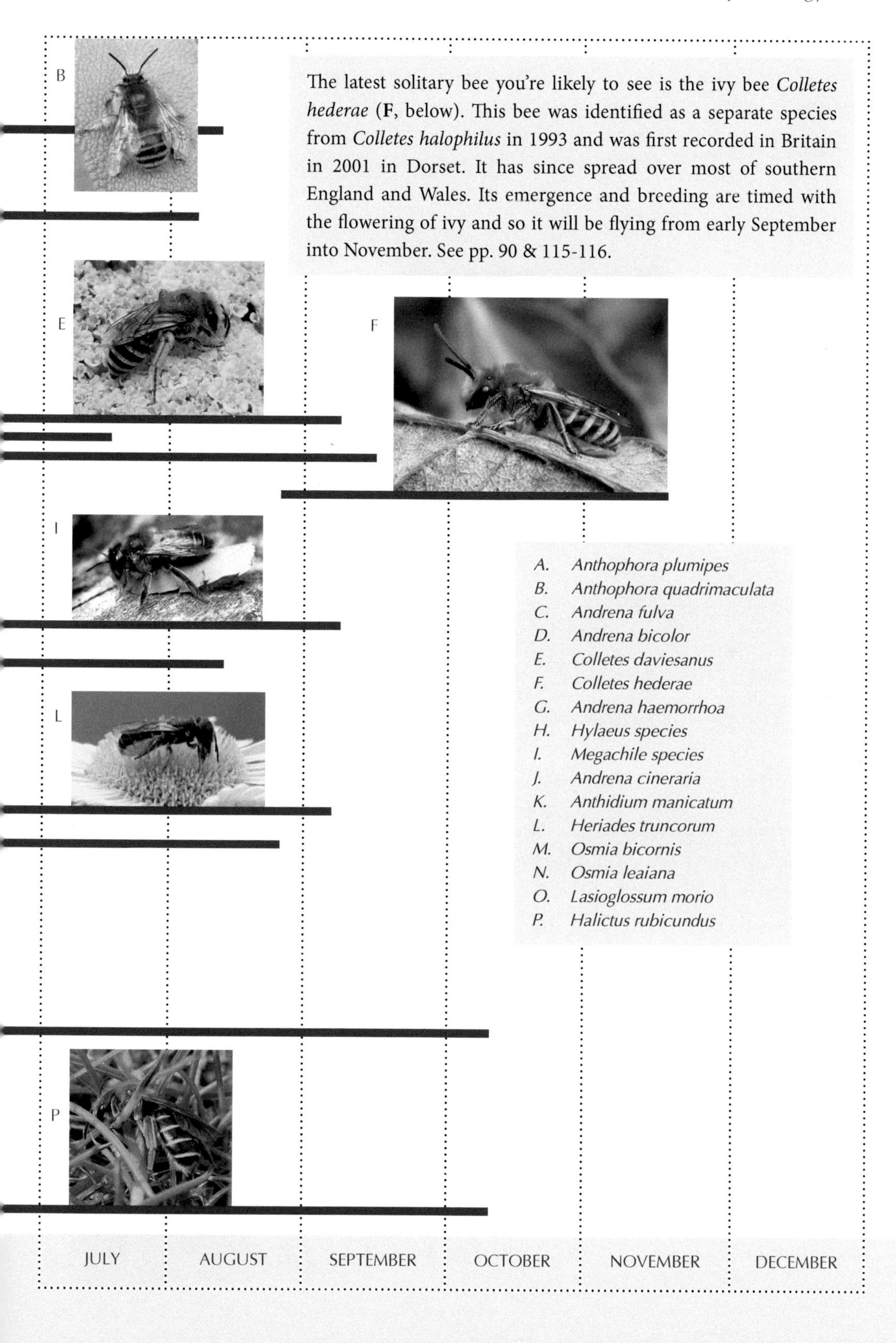

The latest solitary bee you're likely to see is the ivy bee *Colletes hederae* (**F**, below). This bee was identified as a separate species from *Colletes halophilus* in 1993 and was first recorded in Britain in 2001 in Dorset. It has since spread over most of southern England and Wales. Its emergence and breeding are timed with the flowering of ivy and so it will be flying from early September into November. See pp. 90 & 115-116.

A. *Anthophora plumipes*
B. *Anthophora quadrimaculata*
C. *Andrena fulva*
D. *Andrena bicolor*
E. *Colletes daviesanus*
F. *Colletes hederae*
G. *Andrena haemorrhoa*
H. *Hylaeus species*
I. *Megachile species*
J. *Andrena cineraria*
K. *Anthidium manicatum*
L. *Heriades truncorum*
M. *Osmia bicornis*
N. *Osmia leaiana*
O. *Lasioglossum morio*
P. *Halictus rubicundus*

Small cones of freshly piled-up soil in March and April could tell you where the beautiful tawny mining bee is building in your lawn or flower beds. She flies at the same time as the hairy-footed flower bee but she has a short tongue (opposite) and feeds on the flowers of plum, apple and cherry trees, hawthorn and dandelion, rather than on cowslips, deadnettles or lungwort (see p.41).

At the bottom of a vertical shaft (above) she makes tunnels with nest chambers at the end. As she digs a new tunnel, she fills the previous one with the soil. A little bend in the vertical shaft allows her to hide overnight or in bad weather. Her parasitoids (p.112) include the bee-fly *Bombylius major* (above, to the right of the nest hole), which hovers above the nest and flicks its eggs so that they land nearby. When the fly's larva hatches it crawls into a cell in the bee's nest, waits for the bee larva to grow and eats it. The female tawny mining bees above are nesting in bare soil beneath the leaf canopy of *Geranium macrorrhizum*. Another parasitoid fly, *Leucophora*, pursues the female bee to find her nest. The defence of these females was to suddenly dive into the leaf canopy a little distance from the nest hole in order to lose the fly, and then crawl to the nest hidden by the leaves. A third parasitoid, a *Nomada* cuckoo bee, also regularly attends such sites. Where parasitoids are less common, nests are often in large aggregations and can be in an open lawn. Parasitoids are discussed in more detail on pp.103-114.

The female uses her pygidium (visible above right) to compact soil particles on the inner wall of the nest cell and then to spread a secretion from Dufour's gland to seal and waterproof the cell (see also pp.11 & 46). She might make four or five cells per burrow and complete two or three burrows in her relatively short adult life. The effort she puts into avoiding parasites and constructing the nest reflects this low productivity in terms of eggs and young (compared with other insects, such as butterflies, which lay many hundreds of eggs).

The male tawny mining bee is smaller than the female, as with many other solitary bees (but compare with the wool carder bee, p.79). He has very long, curved mandibles that have a hook at the base. He also has long white hairs on his face.

A good bee habitat must provide food, nest materials and nesting sites. There should be a range of flowers throughout the seasons to supply nectar and pollen. Some bees, such as *Anthophora*, feed from many types of flowers and are called **polylectic**. A few seek one flower, such as the ivy bee *Colletes hederae* (p.90) or *Andrena florea*, which feeds on white bryony (pp.99–100) – these are **monolectic**: we need to grow their food plants to attract them. Others use a small group or genus of plants. These are **oligolectic bees**. Some bees (Apidae and Megachilidae) have evolved long tongues, such as *Anthophora*. Others (Andrenidae, Halictidae, Colletidae) have shorter tongues. Tongue length reflects the type of flower the bee favours.

Bee vision is adapted to the blue end of the spectrum, including ultraviolet (short wavelength) light that we cannot see. It is not adapted for detecting red (long wavelength) light. Bees favour blue flowers but will also use red, yellow or white flowers such as cowslip and deadnettle, as we have seen. Some red flowers have ultraviolet patterns in them that we cannot see. These act as nectar guides for bees and other pollinators – radiating lines and patterns that draw the insect down to the nectaries (and past the anthers and pollen).

Anthophora *and other solitary bees seek a range of plants. In addition to cowslip, apples, cherries and strawberries these include:* **1.** *In early spring* Anthophora *feeds on the blue flowers of green alkanet* Pentaglottis *(*Anchusa*)* sempervirens, *a persistent but desirable 'weed'. Lungwort* Pulmonaria *and rosemary are also favourites.* **2.** *Later in spring many bee species feed on native viper's bugloss* Echium vulgare *and its non-native relatives such as giant viper's bugloss* Echium pininana *(opposite, far right), which flowers from May in London. It offers plentiful foraging for bumblebees building nests, late-flying* Anthophora plumipes, *early* Anthophora quadrimaculata *and honeybees. A single stem can host hundreds of bees.* **3.** *Open-flowered roses such as* Rosa moyesii, *here, or wild dog rose* Rosa canina *attract a range of bees.* **4.** *Big purple cardoons in summer (along with knapweeds and thistles) are favoured by* Megachile *leaf-cutter bees (how many can you see?).* **5.** *Lamb's ears* Stachys byzantina *will almost guarantee the presence of the wool carder bee* Anthidium manicatum. **6.** *In summer wild marjoram is loved by many pollinating insects including bees. Here a* Coelioxys *cuckoo bee is just visible.* **7.** *Non-native plants can be as good as wild natives – the South African* Euryops *is used by smaller, short-tongued bees (and many hoverflies).*

What are the best plants to grow?

The Royal Horticultural Society's Plants for Pollinators list and its Plants for Bugs project are among the many online resources available for those keen to plant flowers that will attract bees. There is a comprehensive discussion of planting for wildlife on the Wildlife Gardening Forum website, on the Goulson Lab website and in the book *Plants for Bees*, in which there is excellent advice on plants good for solitary, honey and bumblebees. See Further Reading p.132 and the notes and references on p.166 for more details.

It is not essential to plant native species to provide for bees – many plants shown here are non-native to the UK. Plants indigenous to regions where climatic and environmental conditions are similar to Britain and Ireland will provide for our bee species. Many bees have distributions far beyond UK borders and are adapted to foraging on flora across the Palearctic. *Anthidium manicatum*, for example, currently the only UK species in a genus of over 170, is endemic to Europe, Asia and North Africa. *Stachys byzantina*, one of its favourite plants, is native to Southwest Asia. By growing *Stachys* we are making *Anthidium* 'feel at home', at least a little. Note that *Anthidium manicatum* was accidentally introduced to the US in 1963, where it has become invasive. It joins more than 90 other species of *Anthidium* and is also found in New Zealand and the Canary Islands. *Stachys byzantina*, meanwhile, has escaped from gardens in the UK and is considered a neophyte (a naturalised non-native species).

Rather than consider which plants to grow it is better to ask 'What not to grow?' Choices that will not help bees include trade varieties (cultivars) with double or many-petalled flowers, which provide little or no nectar or pollen.

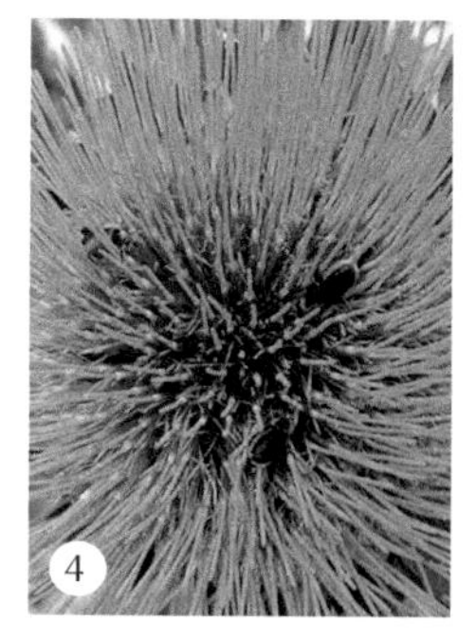

4

5

6

Far right: Echium pininana – *five metres of urban bee forage! This plant is only semi-hardy, however, and needs winter protection unless well sheltered.*

7

The red mason bee is one of the common early spring bees that use 'bee hotels' in gardens. This close-up of the female's head shows her very hairy face and modified mandibles. The projections on the mandibles are perfect trowel-shaped tools and, combined with the pointed part of the mandibles, are used for excavating mud, working it into a ball for transport to the nest site and then for plastering and working with it inside the nest hole.

Red mason bee females are very good at finding areas of damp or clay-type soils. A pond is excellent, of course, but a digging gardener will quickly find these bees hovering around any exposed damp soil. In the photo below left, a female is using a fox paw print next to a pond as the starting point for her mud hole. The females often disappear into deep mud-mines.

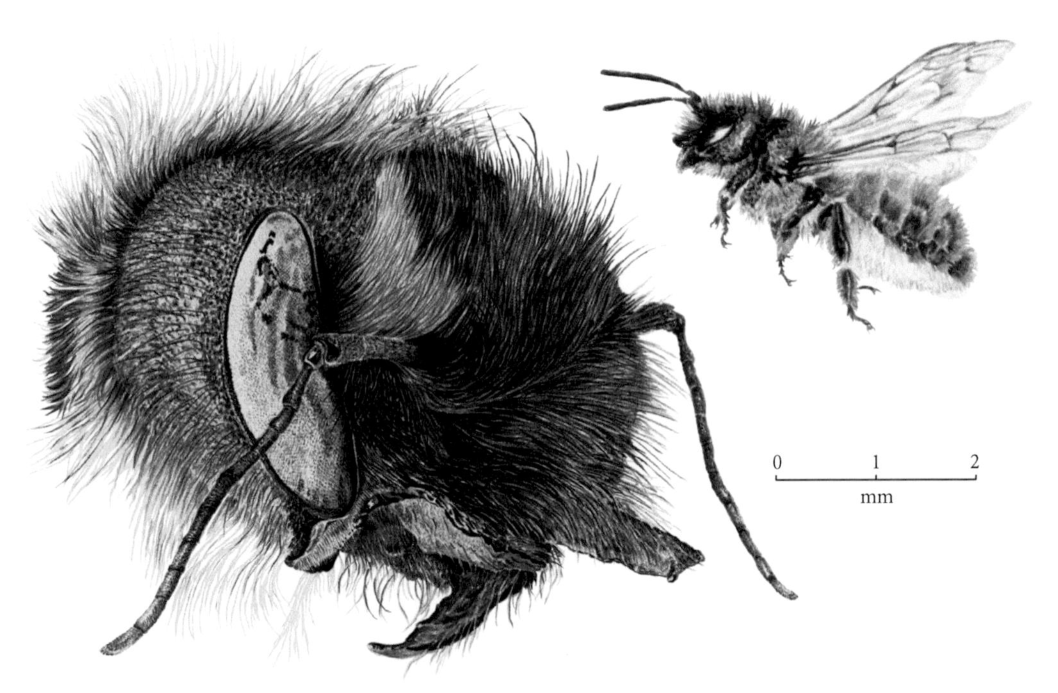

Pollen is carried back to the nest in the long-haired scopa beneath the abdomen. This is typical of bees of the family Megachilidae.

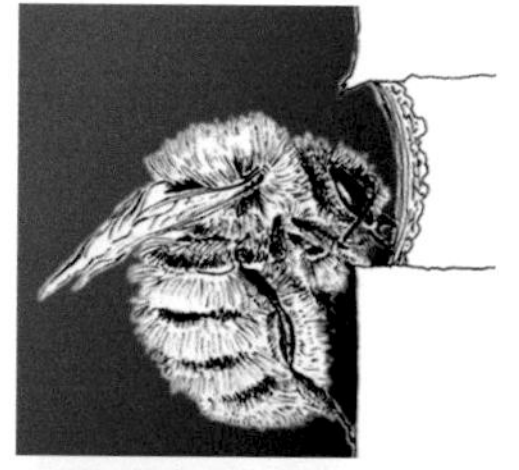

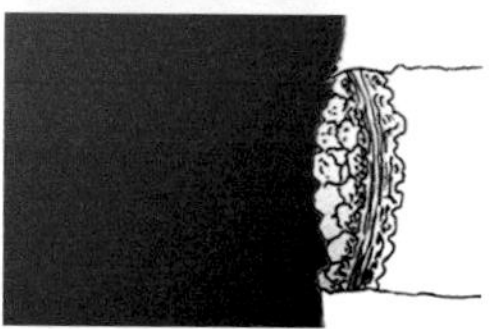

She builds the partitions of the nest in layers. The inner walls are built with a smooth surface facing outwards and a rough surface facing inwards (left). The convex shape of the inner surface will tell the maturing larva which way to orientate itself before pupation so that it faces the entrance of the nest when it is ready to emerge.

The female uses a different technique for the final wall – the nest plug. She sticks blobs of mud to the surface but does not plaster it smooth, because smooth mud would crack and shrink as it dried. Small blobs stick together better, stay stronger for longer and resemble the rough surface of wood. For similar reasons, human builders often roughen the surface of concrete rendering.

Mating in mason bees can take time. Once a male is mounted on a female he will vibrate his antennae above her and use thoracic vibrations to produce a high-pitched hum for around 14 minutes (above left). If his ability to maintain this is accepted, copulation will occur. He will then leave a pheromone signal to show that she has mated.

The bee also has to make the plug thicker than the cell walls inside. The plug has to last almost a year, keeping eggs and growing larvae safe from bad weather and from predators such as birds or wasps. The work is intense and she must also avoid parasitoids during construction (see p.56).

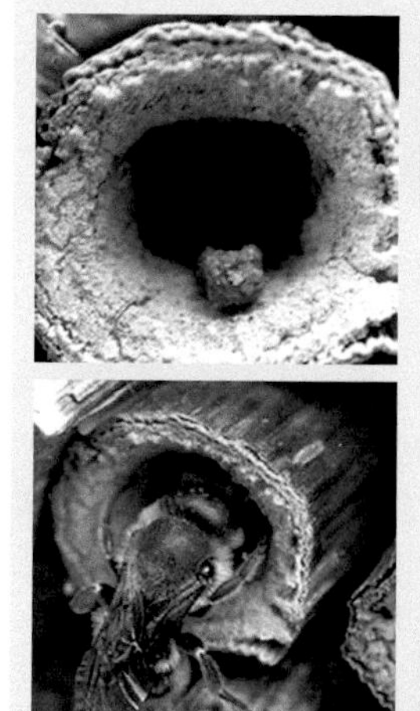

Another species of mason bee (*Osmia leaiana*, left) uses balls of chewed leaf paste as nest material rather than mud. Here a female is working on a nest. Above and below: a nest in a transparent tube.

Bees are opportunists when searching for nest sites but their differing habits can guide our gardening practice. Some *Andrena* mining bees will tunnel into quite compact bare earth, such as alongside pathways, whilst others prefer looser sandy soils or sunny banks. There are many mining bees that use exposed sandy cliffs on the foreshore, such as the glacial till coastlines of East Anglia. To mimic such conditions in the garden can sometimes be difficult but there is one thing to avoid: thick layers of organic mulch such as bark chips, which are impossible to dig a nest hole in, contrary to much gardening advice!

Part of the Trellick Bee Tower (pp.57, 135). It has housed nine bee species, over 250 red mason bee nests, their parasitoids and several solitary wasp species.

Solitary bees that use existing holes – the mason and leaf-cutter bees (and several solitary wasps) – can easily be attracted to human-made structures. Holes in boxes (often called 'trap nests' or 'bee hotels') should be as deep as you can make them – up to 15cm for larger bees. The holes can be hollow tubes, such as canes or plant stems, or drilled into blocks of untreated wood. Tubes should be sturdy and weatherproof (some plant stems will split easily). Shelter from rain – an overhang, if not an actual roof – will increase the longevity of the blocks and pupae survival rates.

Purpose-built rubble roofs – mixed rubble and gravel/sand mounds – are good places for mining bees. These tough environments are less likely to become vegetated. This one has spring bulbs (*Muscari*, alpine tulips and crocus), self-sown *Echium pininana* and purple toadflax *Linaria purpurea*. Even very high roofs can be colonised by *Andrena* species.

Male bees often 'hang out' in bee hotel holes. Left, a male Megachile willughbiella *waits for females in a Japanese knotweed stem.*

Commercially available cardboard tubes with paper roll inners (some are incorporated into the box shown above) need rain protection. If using bamboo canes check inside to make sure they are clear. Drilled holes should be reasonably 'clean' – no splinters or rough edges that could damage the bee; a countersinking tool is useful for this. Both soft and hardwoods are suitable but holes in hardwoods can be cleaner. The back of the hole should be blocked or sealed with no light showing.

A wide range of diameters will satisfy a diversity of bees: 2–4mm for *Hylaeus* bees; 4–8mm for medium-sized mason bees; 8–12mm for larger mason and leaf-cutter bees. Sites should be in full sun for at least half the day and be south or southwest facing. Bee hotels are best located high enough to be above ground vegetation, but bees will use low sites if they are bright and clear. They can look like anything (see Trellick Bee Tower and Bee Shard, pp.135-6) and can be part of garden sculptures or other structures: above and right are designs by students from London Metropolitan University (in the sphere, holes need to be in the lower section only, for shelter). Clay sculptures by Martha Macdonald, below, were very attractive to wool carder bees: the glazed clay is perhaps reminiscent of hot Mediterranean limestone boulders. Boxes of cob (pp.46, 160) can also be constructed using clay loam and barley straw moulded into brick blocks to attract hairy-footed flower bees. As with trap nests, holes in the bricks can be used by several species. See notes p. 167 for more details.

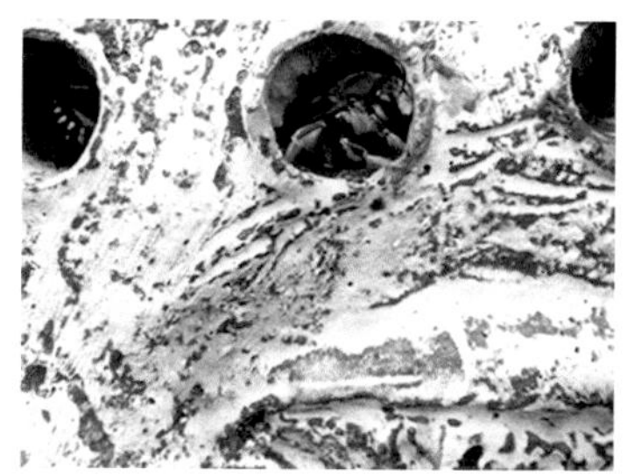

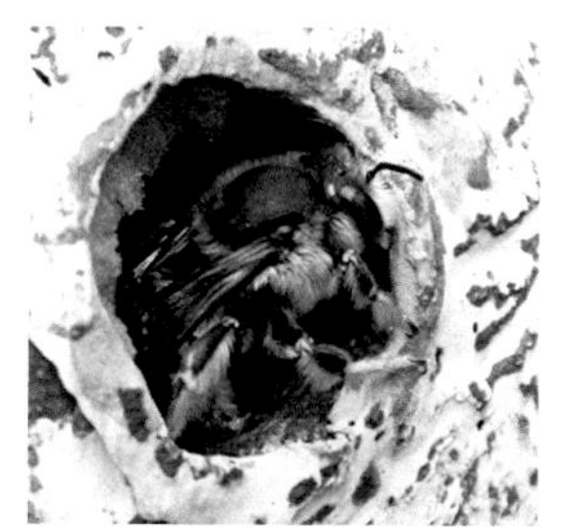

From early June you might spot this striking bee with the same swift and swerving flight as the hairy-footed flower bee. The male is very feisty and will investigate anything new in his territory, including humans. He will often hover in front of you, carefully inspecting, before moving off to more important bee intruders.

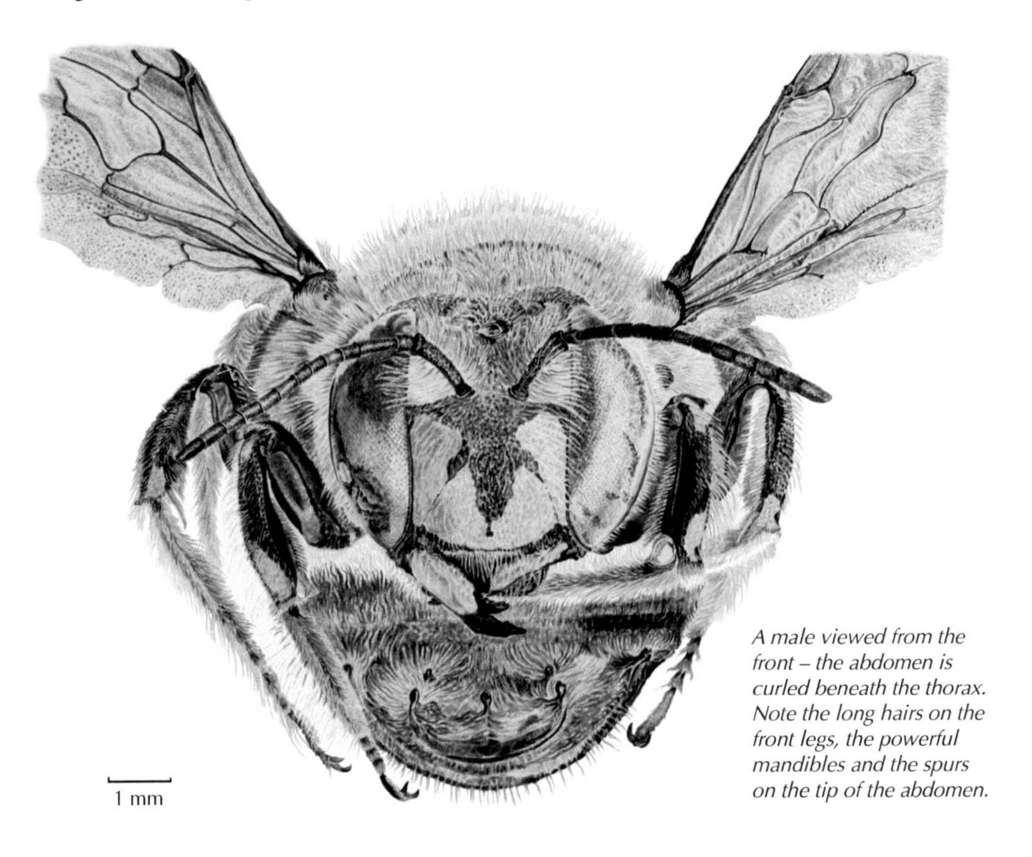

A male viewed from the front – the abdomen is curled beneath the thorax. Note the long hairs on the front legs, the powerful mandibles and the spurs on the tip of the abdomen.

Unlike the male, the female (below) has no spurs. She uses plant hairs as nest material – scraping ('carding') them from leaves and stems (see p.81). The ball of hairs is quickly prepared and is held with the legs beneath the thorax/abdomen whilst she returns to the nest.

The wool carder bee is unusual in the extent of the male's territoriality, and because of this he has evolved to be significantly larger than the female. He has no sting (see pp.65–68) but has evolved sharp spurs on his abdomen. He will patrol and defend against all-comers a favourable patch of plants likely to attract females – attacking bees as large as bumblebees. An intruder is rammed at high speed with the abdomen curled forwards. The mandibles and legs may also grip the intruder. This can cause serious damage, even death, to other bees, though with fellow *Anthidium* males there is usually a 'face-off'. Any female is also instantly pounced on in an attempt to mate. The male covers the female's eyes with the soft, long hairs on his front legs (bottom, right) and possibly uses a pheromone as with *Anthophora* (and *Megachile willughbiella*, p.84). Evolution has altered the behaviour of the female, who will mate more than once. There is evidence that the sperm is not stored in her spermatheca as it is by other bees.

The female, below left, and the male, second from left, both show markings characteristic of the continental subspecies Anthidium manicatum manicatum *(the UK subspecies is A. m.* nigrithorax*). The yellow on the abdomen is more extensive and there are also yellow markings on the thorax. These photographs were taken in central London.*

A knight in bright armour - but with soft-plumed legs and no lance!

If we are providing lots of food and plenty of nest sites, we may as well provide building materials. Of course, a complex garden with lots of habitats and wildness will provide materials we might not even be aware of – for example, bees will find plant resins in places we could never see. However, it's a good idea to include some key elements.

Red mason bees (above and pp.75-76) need a good supply of easily worked mud – so a pond in the garden with soft banks is a bonus. Females are very quick to sense damp soil but nest boxes placed on dry, exposed roofs are unlikely to be used by red mason bees unless we provide damp areas. Wool carder bee males will patrol areas of lamb's ear *Stachys byzantina* (below) – females are attracted to the flowers to feed and to the leaves for nest material. You can see a female *Anthidium* at work and an abandoned ball of wool, below. Other good plants are the giant thistle *Onopordon acanthium* (opposite, bottom left) and Greek horehound *Ballota pseudodictamnus* (opposite, right, with a male *Anthidium*).

Leaf-cutter bees (pp.83-84) need leaves that are easy to cut, carry and mould into their nest holes. The leaves of roses are popular and because of damage to rose leaves these bees have been classified as garden pests in the past. Most useful to them are the thinner, papery leaves of climbing or species roses such as Himalayan musk rose *Rosa brunonii* or wild dog roses, rather than the thicker leaves of some tea roses. Above centre is the *centifolia* rose 'Fantin latour'. *Wisteria* leaves, above left, are also often used but tree leaves rarely so, although false acacia *Robinia pseudoacacia* leaves, above right, are suitable. *Megachile* bees have even used waste plastic as a material.

Plant resin makes a good building material because it hardens (right), is waterproof and is resistant to moulds and bacteria. The large-headed resin bee *Heriades truncorum* often adds grit, sand grains or wood debris to the seal (far right). See pp.95-96 for more details on *Heriades*' nest-building.

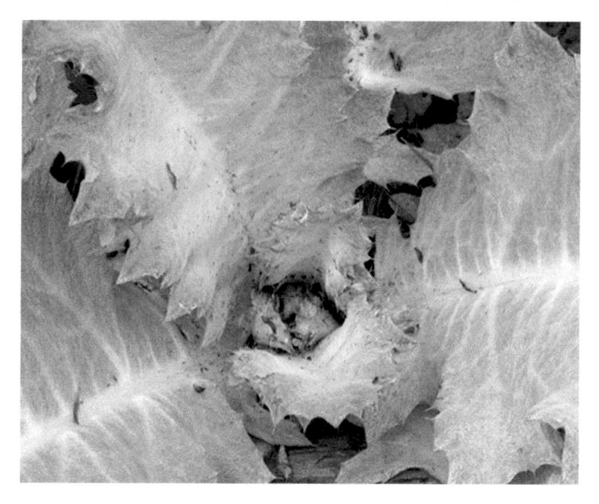

Leaf-cutter bees of the genus *Megachile* are prominent bees in our gardens. The females have a characteristic method for gathering pollen – body often tilted at a high angle, head deep into the flower and mid and hind legs working furiously to shovel the pollen backwards into the hairy pollen scopa beneath the abdomen (*Megachile ligniseca*, below, in favoured knapweed and thistle flowers; see pp.73-4). They can be quite large bees – *M. ligniseca* females have a forewing length of up to 10mm and will use nest holes of 8–10mm in diameter. Leaf-cutters leave their mark in the garden on suitable leaves (see p.82), carrying oval sections back to their nest holes to line the inner walls (bottom and pp.43-4) or circular pieces for cell divisions and the final seal (below, left). Leaf fragments are glued in place with a combination of glandular secretions and sap from the leaves.

True leaf-cutting *Megachile* bees evolved relatively recently – around 22 MYA – but are already extremely diverse and abundant, perhaps because of their well-constructed nests, high reproductive productivity and climatic adaptability. The lining of the nest hole with gathered materials (as in other members of the Megachilidae – *Osmia* and *Anthidium*) was an important stage in the evolution of bees and probably came much earlier – closer to 100 MYA.

Leaf-cutter bee cells that have fallen from a tube in a 'trap nest' show how well the pieces are wrapped together into neat parcels: as tight as a Cuban cigar! The circular ends are cut perfectly to size, although sometimes the female will make mistakes (as at the top of the post, left), perhaps when her behavioural 'algorithm' is interrupted by weather, predators or parasites.

Hairy-foot - smelly-foot!

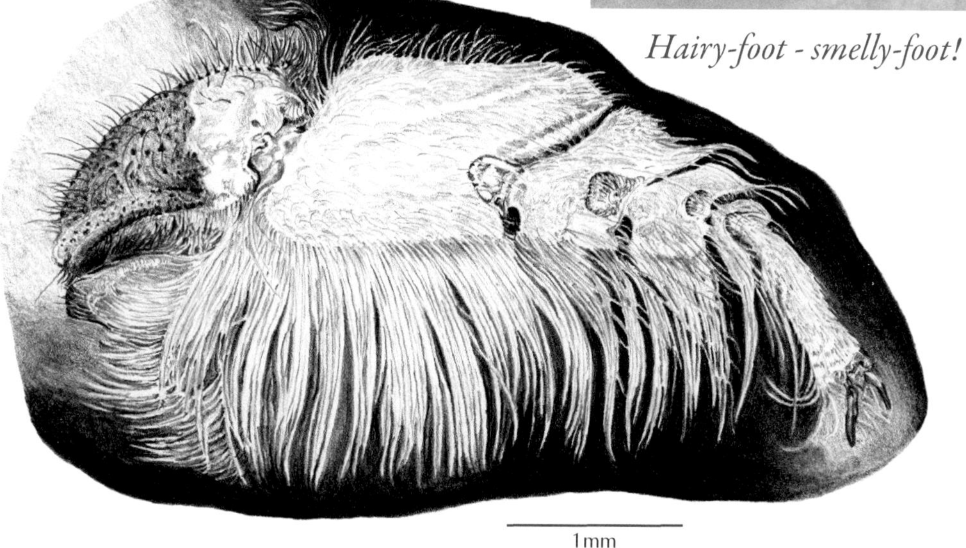

The male of the smaller *Megachile willughbiella* has broad, swollen-looking front tarsi with long fringes of pale, creamy hairs; with his substantial mandibles this gives him a distinctly pugnacious air! During mating the tarsi are used in a similar way to the hairy legs of *Anthophora* or *Anthidium*. He also has a deep groove on the underside of each basitarsus (not visible in the drawing above), which has scent glands. The grooves trap the female antennae whilst the hairy fringes cover her compound eyes. His mid legs trap her wings and his hind legs hook beneath her abdomen: she is blindfolded, perfumed and immobilised! When he sits out poor weather or overnight in old nest holes he will often hold his forelegs across the entrance – seemingly for protection (right).

The Wildlife Garden at Roots and Shoots in Lambeth, central London has been 'wild' for nearly 40 years and has been managed with bees in mind since 1999 – the majority of the bee species in this book have all been seen there at some point. Visits can be made by appointment. Before the COVID pandemic the garden was used for educational programmes and visits: these began again in late 2023. It is around a half-acre in size with a half-acre of additional gardens and another half-acre of public open space alongside (below right). There is a small orchard of local apple varieties, which was established in 2000. Over 30 species of bee have been recorded at the site but each year there are some that remain unidentified. This suggests a welcome biodiversity in the inner city.

The importance of cities for bees and other pollinators has been known some years. However, it can be a small and vulnerable diversity, especially when suitable habitat is scattered and separated by hostile built environments and subject to significant pollution. Even with careful oversight, changes to the structure of a small garden – for example maturing trees – alter its suitability for various species. The Trellick Bee Tower (p.57), for example, has had to be moved away from the growing shade of an oak canopy. Habitats at Roots and Shoots include meadows, ponds, mature trees, coppiced hazel, hedges, mixed borders, a rain garden and planted roofs. The photos below show roofs on which grape hyacinth *Muscari*, red valerian *Centranthus*, *Sedum* and thrift *Armeria* are growing – all excellent for bees, flies and moths.

There are now many wildlife gardens to visit around the UK. These include the RSPB's Flatford Wildlife Garden and Frampton Marsh, Kent Wildlife Trust's Tyland Barn and other Wildlife Trust sites, the Centre for Wildlife Gardening in south London and the Johnston Terrace Garden in Edinburgh. Some city parks are also being managed with wildlife in mind. Bee-friendly grassland is encouraged in London's Royal Parks, and the National Botanic Garden of Wales is undertaking research and education programmes for pollinators, with meadow planting a feature of its grounds.

The charity Buglife runs the 'B-lines' project and Friends of the Earth have been campaigning for bees with their Bee Cause programme since 2012. They have many suggestions for getting involved. The Wildlife Gardening Forum organises conferences on garden science and gardening techniques and has a comprehensive website. Membership is free and conferences are affordably priced. All of this reflects the growing popularity of wildlife gardening. The RSPB, Wildlife Trusts, RHS and the Natural History Museum provide extensive general (though not necessarily bee-specific) guidance including at events, and have examples of gardens in practice. See 'Further Study, p.131 and notes, p.160 for websites and links.

Above left is the Wildlife Garden at Roots and Shoots and above right is RSPB Flatford Wildlife Garden in Suffolk. Both demonstrate wildlife gardening techniques and planting. Flatford has a wonderful pond and a bee cob wall. Below: although somewhat larger than your average garden, Plantlife's Three Hagges Woodmeadow south of York shows what can be done for biodiversity on former agricultural land – excellent lessons for our urban fringe or, indeed, for city parks.

Bees in the genus *Hylaeus* are unusual in several respects. They are part of the family Colletidae with the genus *Colletes* (p.89) and are often called 'yellow-faced bees' because the males of all but one of the 12 British species (*Hylaeus cornutus*) have yellow faces. They are all small, have forewing lengths of between 3.5 and 6mm, and are virtually hairless. This makes them appear very wasp- or ant-like and they are undoubtedly often overlooked in gardens for this reason. They also have no pollen scopae or pollen-gathering hairs, denying their very 'bee-ness'. They were once thought to represent a 'basal lineage' in evolutionary terms but we now know that the Colletidae are relatively recently evolved and that the hairlessness is a 'derived' feature – in other words, they lost their hairs in their more recent evolutionary past. In 2013 scientists Sophie Cardinal and Bryan Danforth proposed in their 'family tree' for bees that Hylaeine bees first appeared in the early Miocene period 20 to 30 MYA (pp.122–123). *Hylaeus* is a very large genus with up to 700 species worldwide. It is most diverse in Australia where it probably first occurred. It is also, amazingly, the only bee genus present in the Hawaiian archipelago, having reached there only in the last half million years or so (probably from Japan or east Asia, perhaps in nests in floating timber). In that relatively short time it has radiated into 63 species (11 new species have been found since 2000) occupying all types of habitat on the islands and becoming the most important pollinator for much of the native flora. Five of the Hawaiian species have become cuckoo bees; cleptoparasitism is otherwise unknown in the Colletidae. However, because of habitat destruction and declines in native flora, seven of the Hawaiian species are now listed as endangered.

Nests are built in hollow plant stems or holes in timber (such as beetle larvae emergence holes or old nests of other Hymenoptera) – or in our bee hotels. Megachilidae is the only other genus of bees to do this. Females (below) work diligently to seal the nest hole, rotating the body around the hole whilst pasting the secretion used for the seal with an upwards motion of the head. They are often known as 'plasterer bees'.

Yellow-faced males, like other solitary bees, patrol suitable feeding and nesting sites hoping to mate and frequently shelter in empty holes (right). Males of the larger species can be aggressively territorial around flower patches, as *Anthidium* males are, and have small projections on the third abdominal sternite for use in skirmishes.

As with *Colletes* (p.90), the nests of *Hylaeus* are lined and sealed with glandular secretions spread and applied with the short, forked glossa. In *Hylaeus* it appears that the secretions are mainly from the thoracic salivary gland. The secretions solidify during nest-building to resemble spider silk. Secretions from Dufour's glands (p.12) are probably mixed with these and with fungicidal and anti-bacterial secretions from the mandibular glands to ensure the survival of the nest contents inside the cellophane-like lining. *Hylaeus* must carry pollen in the crop, mixed with nectar. Females will increase the concentration of nectar by 'blowing bubbles' (right) to evaporate some of the water content (see also p.93) but the food mass in the nest hole is still of low viscosity, making the waterproof lining even more essential. These images show *Hylaeus* females apparently gathering material from the waxy surfaces of *Fremontodendron* petals and from the nectaries. However, their small size and lack of hairs makes them poor pollinators for many plants.

Colletes daviesanus, (below and opposite, top), is quite a frequent bee in urban gardens but unlike its 'sister' genus *Hylaeus*, *Colletes* bees are true miners. The glossa of both *Hylaeus* and *Colletes* is short, like *Andrena*, but bi-lobed – forked at the tip – rather than pointed or long and slender as in *Anthophora* (below, the head and mouthparts of a male *Colletes daviesanus*). Bi-lobed glossae are used by females to apply and spread the glandular secretions used in nest-building. In *Colletes* these are primarily from Dufour's gland, probably mixed with mandibular/ salivary gland secretions (see *Hylaeus* p.88). The resulting nest cells can appear to be wrapped in thin, clear cellophane, which waterproofs the nest to such an extent that it can survive flooding by river or tidal water. The wrapping also stops the runny mixture of pollen and nectar soaking into the surrounding soil.

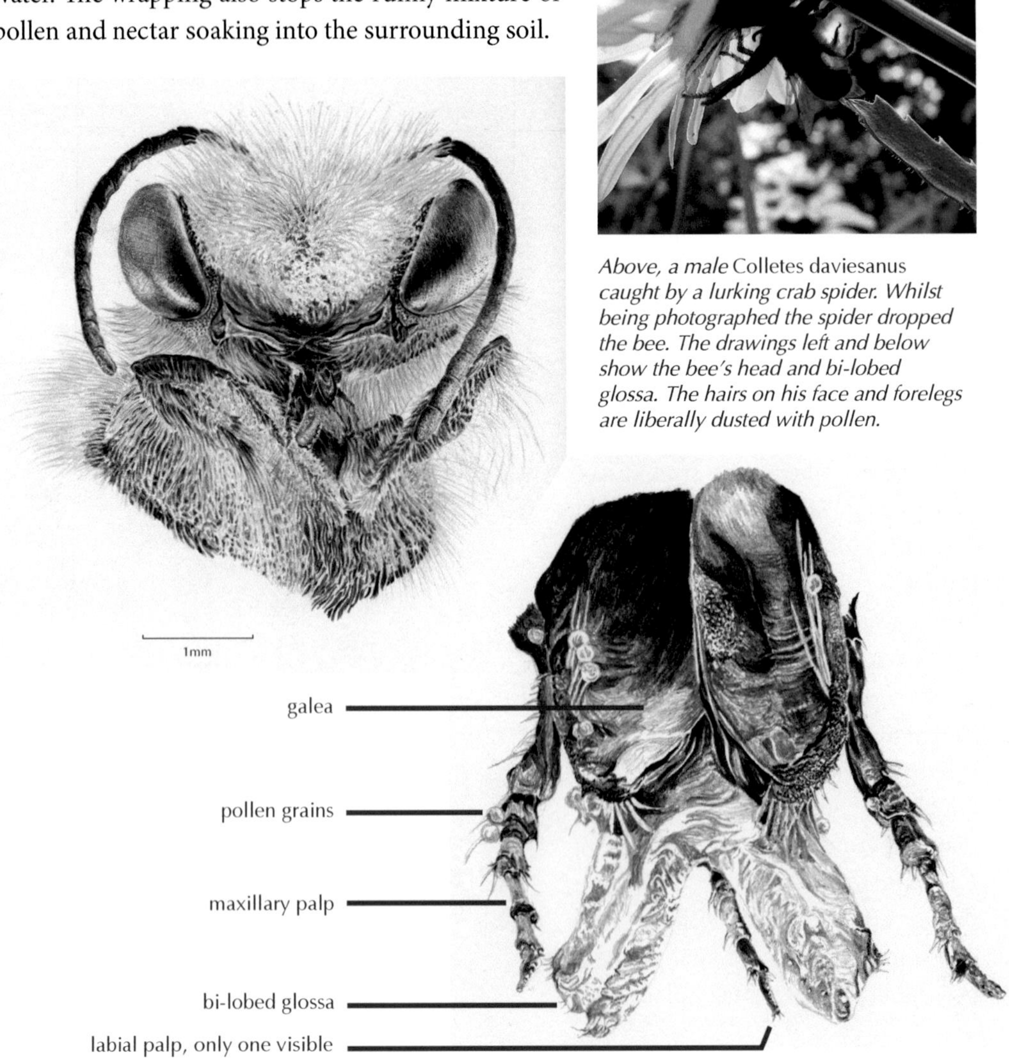

Above, a male Colletes daviesanus *caught by a lurking crab spider. Whilst being photographed the spider dropped the bee. The drawings left and below show the bee's head and bi-lobed glossa. The hairs on his face and forelegs are liberally dusted with pollen.*

The ivy bees *Colletes hederae* pictured below were making burrows in chalk cliff debris near Margate on the Kent coast. Although most nests here were in the gritty soil cover above the cliffs, or in landslips, one female made valiant efforts to excavate in the chalk rubble. *C. hederae* was only separated from two related species in 1993 and only recorded in mainland UK in 2001. Its rapid spread from the south and east has been well documented and is described on pp.115-116. It is now well established in London and the southeast and most sunny clumps of ivy are well used by the bees in September and October, the only limitation perhaps being suitable nest sites nearby. The bee emerges late in the season to coincide with ivy flowering (below, top right) and pollen for the nest is gathered almost entirely from ivy. Male bees, below bottom left, emerge earlier than females. Because the ivy was not quite ready they were nectaring on Bristly oxtongue *Helminthotheca echioides*. Nests are built in late autumn and, unusually, larvae overwinter when they are half-grown. The waterproofed cell lining keeps food supplies edible whilst they grow. They pupate the following spring. The way in which the cell lining has evolved enables this late flight and nesting time.

Andrena is the largest genus in the UK. There are 68 species, many of which might be described using the birdwatchers' term 'little brown jobs': species ID can be difficult! Globally there may be 1,500 species. These are subdivided into 94 subgenera, including the 'mini-miners' (*Micrandrena*). The group also contains some striking species, which are more easily seen and identified. Some we have met already: *A. fulva, A. bicolor, A. haemorrhoa, A. cineraria* (below) and *A. nigroaenea* (bottom left). Nests are built into bare ground, from level to steeply sloping, and sometimes in soft mortar or wall cavities. The bees are all solitary, though some will nest in large aggregations and possibly communally (see p.58).

The distinctive ashy mining bee (above) occurs over much of Great Britain and Ireland and is probably becoming more common in the south. It can be seen in urban and rural areas. There has been a small aggregation, for example, in the gardens of Buckingham Palace (where its cleptoparasite *Nomada goodeniana* (p.109) is also found). The images above are from a rural garden in Northern Ireland. Here the ashy mining bee can face poor weather in early spring when it emerges and so will forage on whatever is available – daisies (top left), for example, and early flowering wild cherry. *Andrena nigroaenea* and *Andrena nitida* (below, left and right) often nest singly, the former digging into more loamy soil such as silted crevices in garden walls, the latter on level ground. They are both common in gardens across southern Britain. *A. nigroaenea* is also found in much of Ireland. Both images below show the well-developed scopae on most parts of the hind legs and the propodeum: many *Andrena* carry large, loosely packed loads of dry pollen and so are excellent pollinators.

The spring-flying *Andrena* species are very useful fruit pollinators: the orange-tailed mining bee *Andrena haemorrhoa*, for example, is a reliable pollinator of apples. It is often seen in urban areas and is common and widespread across Britain and Ireland and up to the far northwest of Scotland.

Many *Andrena* species are hosts for *Nomada* cuckoo bees (see p.109). *A. haemorroha* hosts the striking and almost equally widespread *Nomada ruficornis*, whilst *A. clarkella* (lower left) supports *N. leucophthalma*, which is pictured below carefully searching for its host's nest. It then dug into a sealed burrow in the banks of a river in Northern Ireland. A female *A. clarkella* may well temporarily disguise her nest during construction and provision but the *Nomada* will watch carefully and dig quickly.

Andrena clarkella can emerge from late February. It is oligolectic on willow *Salix* species. *A. clarkella* is also widespread, being found in the far northwest of the UK and across Ireland. It can be locally common and occasionally forms large aggregations.

The family Halictidae is represented in Great Britain by *Halictus, Lasioglossum* and the cuckoo bee genus *Sphecodes*. Two smaller genera, *Dufourea* and *Rophites*, are considered extinct. *Halictus* and *Lasioglossum* are characterised by a groove (or rima) along the female's fifth tergite and so are called furrow bees (p.61). We met *Halictus rubicundus* (below, centre) in the discussion of sociality. In the upper image a female *Halictus tumulorum* is drawing out nectar from her crop onto the mouthparts. This exposes it to air, reducing the water content of the nectar by evaporation (compare with the female *Hylaeus* on p.88). Her short, pointed glossa is visible. These latter two species are the most likely *Halictus* to be found in gardens. Some *Halictus* and *Lasioglossum* species are difficult to distinguish. Pale bands on the hind edges of each tergite identify end-banded furrow bees (left, centre). Although most British species are small, *H. rubicundus* is around the size of a small, slender honeybee. Rare species, such as the giant furrow bee *Halictus quadricinctus*, are noticeably larger.

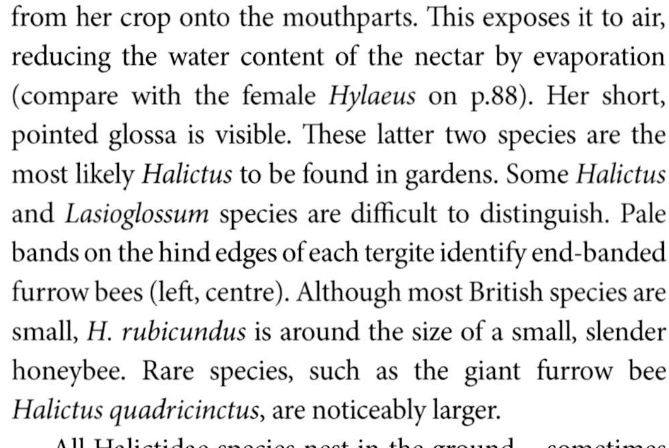

All Halictidae species nest in the ground – sometimes on cliffs or in old walls – and are often seen in large aggregations. There are both solitary and primitively eusocial species within the family. The large European species *H. scabiosae* (found only on the Channel Islands in Britain) has extremely flexible lifecycles: some nests have single females with some worker daughters only, others comprise groups of related females. The whole family is of much interest to evolutionary science (see pp.61-63).

Most species collect pollen using scopae on the hind legs though some also use hairs on the sternites beneath the abdomen (left, centre). There are eight species of *Halictus* and 34 species of *Lasioglossum* in Great Britain and Ireland (and around 300 species of *Halictus* and between 1500 and 1700 *Lasioglossum* species worldwide; *Lasioglossum* is the largest bee genus in the world).

There are 17 *Sphecodes* cuckoo bee species in the British Isles and more than 300 globally. The photograph, left, was taken in the same area as that of the female *Halictus rubicundus*, mid left, in early spring and probably shows a female *Sphecodes gibbus*. The latter is a cleptoparasite of *H. rubicundus* and of several *Lasioglossum* species, including *L. calceatum* (opposite, far right). However, they are difficult to study and the hosts of many *Sphecodes* species are not clear.

Unlike *Halictus* species, *Lasioglossum* bees have lighter bands of hairs at the base of each tergite (i.e. nearer the thorax) and so are called base-banded furrow bees. The female *Lasioglossum morio*, above, is foraging on white bryony *Bryonia dioica*, the only member of the cucumber family native to the UK. The plant is attractive to several smaller solitary bees, including the monolectic specialist *Andrena florea* (p.99), but can be invasive: it is only suitable for larger and wilder wildlife gardens. *L. morio and L. calceatum* are frequent and widespread species that are often found in gardens. The latter reaches most of Scotland and Ireland. Both have a long flying season and a primitively eusocial lifecycle. Males have distinctively long antennae. The male *L. morio*, below, left, is on the tiny flowers of wild fennel. Male *L. calceatum*, right, have a variable amount of red on the abdomen. This one is on a cranesbill flower.

As with other members of the Megachilidae family (e.g. *Osmia, Anthidium* and *Megachile*) the female *Heriades* has a prominent hairy scopa beneath the abdomen, which she vibrates when foraging. This, and the use of the hind legs, (below) collects pollen very efficiently. Although there are over 100 known species in the genus, only *Heriades truncorum* breeds regularly in the UK, although *Heriades rubicola* was thought by researchers to be breeding in Dorset in 2017 and could be under-recorded. *H. truncorum* is restricted to the southeast, has become more frequent in recent years and may be moving northwards.

The females favour yellow-flowered plants in the family Asteraceae, such as ragwort, common fleabane (above), hawkweeds and sowthistles. Meadow seed mixes often contain these attractive flowers, and North American plants such as *Helenium*, found in 'prairie mixes', will also attract *Heriades*. The species is small compared with other members of the family (though females vary in size) and emerges in summer. It looks quite dark and has sparse tawny hairs on the upper surfaces of the thorax and abdomen and narrow white bands along the rear margins of the tergites. The head appears broad and chunky compared with the bee's relatively long, narrow abdomen. The surface of the body is marked with 'dimples' (it is 'coarsely punctate'), which are visible in some of the photos below and opposite.

The nest is built in holes in timber (such as beetle larvae emergence holes) or in hollow sticks and tubes between 4 and 6mm in diameter. Bee hotels have become very attractive to this species. The cells and entrance are closed with plant resin often from pine trees. Where there are no pines nearby, resin from chestnut trees or softwood timber cladding on buildings or fences could be an alternative source. Tree resin is an excellent building material – it sets very hard and is resistant to moulds and bacteria (p.82) though collecting and transporting it could cause problems – hence the many fossilised bees in amber

*Above: profile of a female showing her pollen scopa and the 'beard' of short hairs beneath the mandibles that assist in the carriage of resin. The male, right, has pale facial hairs and a distinctly downward curving tip to the abdomen (**clavate**). This is best seen from the side.*

(pp.47-50; this indicates the antiquity of its use by bees). The entrance plug is reinforced with materials such as grit (opposite, bottom right) or splinters of wood. This strengthens and helps to disguise the nest entrance. Although Megachilid bees are solitary nesters, the two females pictured below right are sealing the same nest: they worked at it without signs of conflict but it was unclear whether one was the daughter of the other from earlier in the season and acting as a 'support worker' or whether the collaboration was due to confusion. If the former, the behaviour could indicate low levels of cooperation; if the latter, it might suggest low levels of aggression. The rare cuckoo bee *Stelis breviuscula* uses *H. truncorum* as its host and does look very similar.

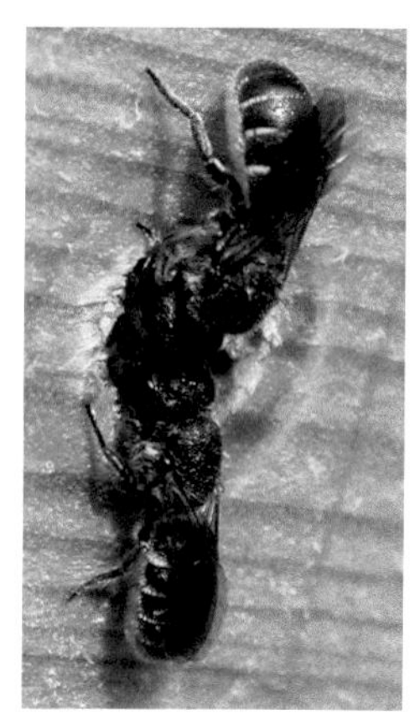

Not long after the disappearance of the last female *Anthophora plumipes*, by late May or early June, you might see another beautiful *Anthophora* – the four-banded flower bee *Anthophora quadrimaculata*. It is smaller than the hairy-footed, and the male (below and opposite, top) is even more swift and swooping in his patrolling of suitable flower patches. He loves lavender but will also be found on black horehound, deadnettles, *Echium* and *Verbena bonariensis*. If there is more than one male around the flowers they will put on a spectacular flying display.

The species is restricted to the southeast of England, its most northerly points currently being sites in Norfolk and Northampton, but it is frequently seen in Greater London. It may be consolidating its populations. Lavenders planted at Waterloo East station in 2017 were quickly used by a male, and exactly a year later by three males (see pp.167-8).

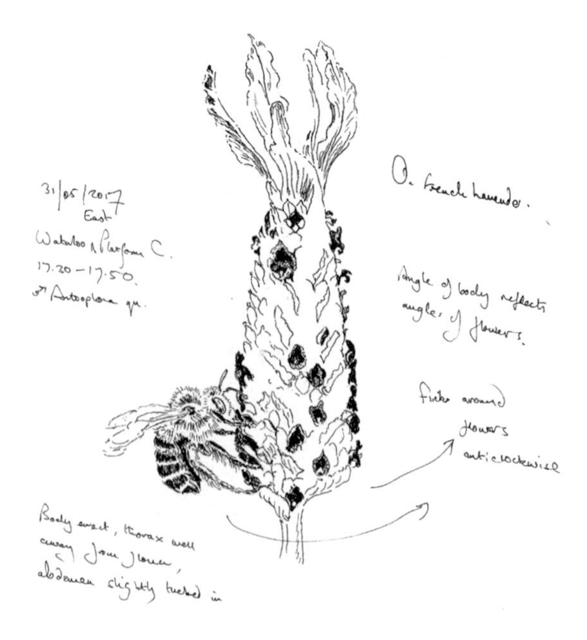

Although its eyes are not quite as green as those of another relative, *Anthophora bimaculata,* they are still striking (opposite left, the compound eyes can just be seen against the green of the sage leaf). The yellow face of the male is broken by two black marks below the antennae – this distinguishes him from the male *A. bimaculata* (p.129). The face of the female is dark (right, and below right feeding on viper's bugloss). She builds nest holes in sunny, sandy banks, on soft cliffs or in walls, sometimes in small aggregations. The very rare cleptoparasitic bee *Coelioxys quadridentata* has been known to use *A. quadrimaculata* as a host. There are five species of *Anthophora* in Britain but over 450 worldwide. The genus probably arose in the Mediterranean region but it is also well represented in the Americas, though it is absent from Australia and Madagascar.

The strategies plants employ to encourage bees to transport their pollen and those used by bees to collect it for their larvae are complex. Bee adaptations have made them reliable flower visitors and the adaptations in flowers to exploit these visitors can sometimes seem bizarre (for example, some tropical orchids provide scented oils with which male euglossine bees can create a species-specific, female-attracting scent bouquet. Males have enlarged urn-like structures on their hind legs and other structures that can absorb and distribute these oils). Pollen collecting can also become specialist – such that some species will only collect from a single species of plant. These are monolectic (see also p.73). There are few strictly monolectic bees in Britain – the ivy bee *Colletes hederae* (p.90) and the bryony mining bee *Andrena florea*, are examples. In Britain *A. florea* is only known to take pollen from white bryony, though it will take nectar from elsewhere (just as *C. hederae* will also stray from ivy). In Germany, however, bryony mining bee females also collect pollen from the closely related *Bryonia alba* plant. This does not occur in Britain but if it did, the mining bee would probably not be monolectic, but **narrowly oligolectic** (collecting pollen from a narrow range of related plants) instead. The plants themselves are not so fussy – many species of bee are attracted to white bryony, including *Lasioglossum* species (p.94) and honeybees. Late-flowering ivy is definitely not dependent on the ivy bee: it is an important forage plant for many pollinators, including flies and butterflies.

The female Andrena florea *has well-developed pollen scopae including the floccus. She forages for pollen solely from the small white-and-pale-green flowers of white bryony, a scrambling, climbing plant whose long coiled tendrils cling and support it through the hedgerow. The males, opposite, are settling in to roost in flowers for the night after a hard day milling around hoping to mate.*

Whilst the specialism in these cases seems to be of the bees, they have no specialist structures with which to select only bryony or ivy pollen. It may be that the larvae gain an advantage from certain pollen but given the wide range of insect species using these plants this is unlikely. A behavioural advantage seems more likely to have caused the evolution of these specialisms: both plants are rampant climbers in hedgerows providing a mass supply of pollen for enough time to make nest provisioning easy for the bee. The book *Solitary Bees* by Ted Benton discusses in some detail the research on the complex networks of pollination systems.

I planted white bryony at Roots and Shoots in 1999 for its botanical interest – it is the only member of the cucumber family, Cucurbitaceae, native to the British Isles – but had to wait until May 2019 for *Andrena florea* to turn up. White bryony is **dioecious** (hence its Latin name *Bryonia dioica*), which means that it carries pollen only on a 'male' plant and has ovaries on separate 'female' plants. Other members of the family also produce separate male and female flowers – but on the same plant (e.g. zucchini). White bryony has a very large tuberous root which if split reproduces. The roots, vegetation and small red berries of white bryony are highly toxic and will cause severe diarrhoea if eaten (though they are also very distasteful). It has in the past, perhaps dangerously, been used as a herbal medicine.

The appearance of a new bee in the garden is exciting and suggests that if you were only out more often you might find more species still! How they end up in relatively isolated spots of suitable habitat in urban areas is intriguing. The two species here appeared at Roots and Shoots in central London: it is possible that they followed the Thames from habitats further east along the Essex and Kent coastline. Below, on the lovely pink rose 'Fantin Latour', is *Andrena nigrospina* or *A. pilipes* – it is unclear from the photos which species this is. She foraged for 12 minutes on the rose in 2008. *A. nigrospina* is more frequent inland but the movement westwards from the coast of a female *A. pilipes* is entirely feasible: the bee has been recorded on the Greenwich Peninsula (2016), for example. The pale hairs on the hind tibia in the female on the rose are hidden by the pollen but are visible in the female foraging on the lesser knapweed in the Cellini Orchard at Roots and Shoots in 2014. *A. pilipes* has two flight periods, April/May and July/August (when these females were seen), and is more frequent than *A. nigrospina*. It prefers to nest in near-vertical sandy slopes – hard to come by in that area, so the nest site remains a mystery.

This other rarity is the hairy-legged mining bee (also called the pantaloon bee) *Dasypoda hirtipes*. The female has extravagantly shaggy hind legs with long, elegant yellow hairs, but the pollen scopae are not restricted to her tibiae: as in some *Andrena*, she has long hairs on other joints of her legs, the coxa, the propodeum and on the forelegs and mid legs (see *Andrena haemorrhoa*, p.24). This female was feeding on field scabious in the Wildlife Garden at Roots and Shoots – though this species generally collects pollen from the yellow Asteraceae such as ragwort, fleabane, hawkbit, cat's-ear or sowthistle – so she may have been foraging only for nectar. A hairy-legged mining bee had also been recorded on the Greenwich Peninsula in 2016.

This bee builds her nest in sandy soils, using those long-haired hind legs to brush sand backwards away from the nest hole, leaving a characteristic 'fan' of waste sand (bottom right). Nests have been recorded in very large aggregations and in densities of up to 21 nests per square metre. The tunnel can be more than a metre deep in suitable sites. The Dufour's gland is absent, so no cell-lining secretion is produced. Instead, to protect the food store and larva from damp and mould the bee sculpts the thick pollen/nectar mix into a ball supported on tripod legs. This lifts it clear of the cell floor and walls (bottom left).

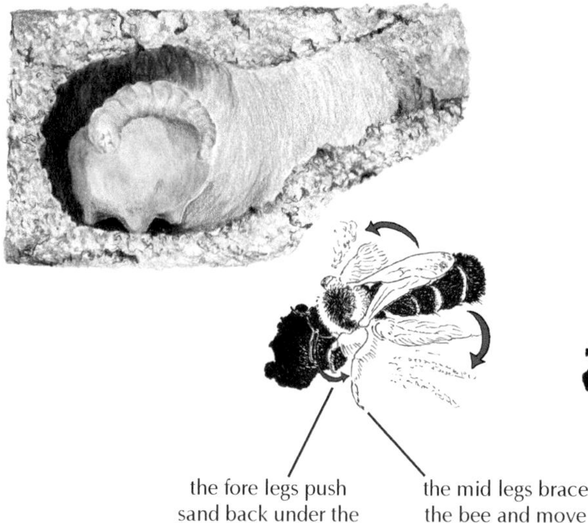

the fore legs push sand back under the thorax and abdomen

the mid legs brace the bee and move it backwards

the hind legs arc forwards from the abdomen towards the head then back, repeatedly brushing sand outwards in fans

Bee nests are attractive food sources to other animals (see pp.67–68). They contain a pollen-nectar food supply and the protein of the eggs and larval bees. The evolution of the Hymenoptera has produced a large number of species that exploit other ants, bees or wasps as food sources for their own young. They may be:

- **predators**, catching and killing prey as food and to stock nests
- **parasites**, using another species as a food source but without killing the host; these may be harmless (when they are called **commensals** or **inquilines**)
- **parasitoids**, in which the lifecycle appears to be parasitic but where the host is eventually killed
- **cleptoparasites**, that consume a stored food supply, often without directly destroying the host larvae. These are sometimes called **brood parasites.**

Bee nests can also be attacked by non-hymenopteran predators, parasites and parasitoids such as flies, mites, beetles and spiders, and large predators such as birds (bee-eaters, nuthatches, tits, woodpeckers) or mammals such as badgers. On the following pages hymenopteran cleptoparasites, parasitoids and predators are considered first, followed by the cleptoparasitic cuckoo bees and finally troublemakers from other groups.

The female cleptoparasitic wasp *Gasteruption jaculator* has a remarkable body form and is highly adapted to the many challenges of its lifecycle. The male, below, is equally strange looking, though he lacks the long ovipositor. He is foraging on a *Euphorbia* flower.

To find a host nest, the female *Gasteruption* hovers above one nest hole then another, using her sensitive antennae to detect activity. She inspects the nest by flicking her antennae across and into the hole, before unsheathing her ovipositor (opposite top left and right). She adjusts her body and ovipositor before carefully attempting to penetrate the seal protecting the nest hole (opposite, centre). Once through, the ovipositor is thrust downwards (opposite, left). If the hole is not completely filled with sealed bee cells she will insert her whole abdomen to reach further (bottom, centre). The photos, bottom right, show the holes in two *Hylaeus* nest seals where *Gasteruption* has laid. The left-hand nest also has two tiny test holes.

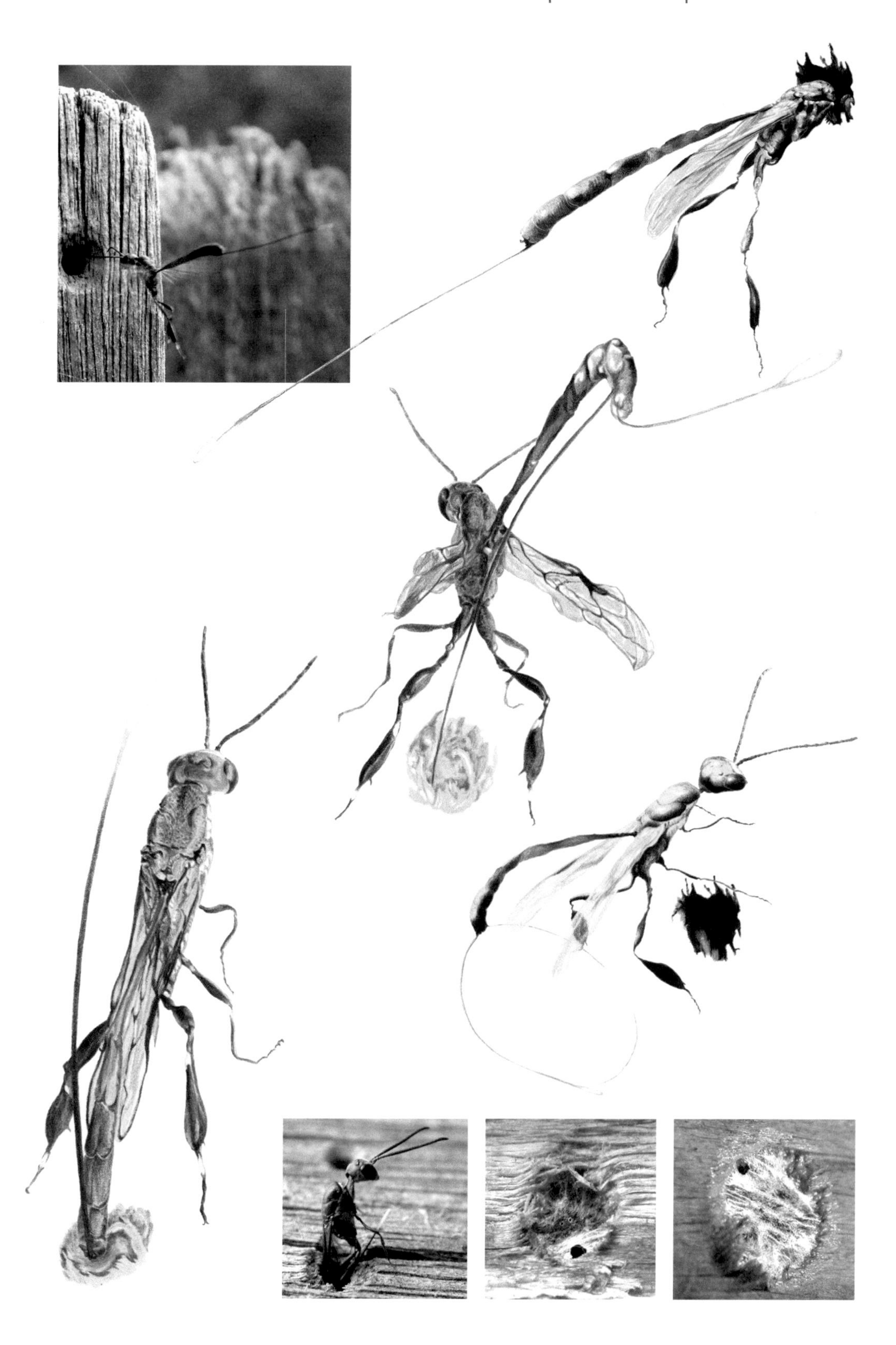

The cuckoo lays her eggs in the nests of smaller birds, which rear her young whilst she migrates south. This is brood parasitism, though in the case of the common cuckoo and a host such as the reed warbler, none of the warblers' young survive the parasitism. When a crow's nest, however, is parasitised by a great spotted cuckoo *Clamator glandarius*, a bird found in Africa and southern Europe, some of the host young survive and nests may be more likely to succeed than those not parasitised. This may be because the cuckoo chick emits a noxious anti-predator smell, although this is not known for sure. Many species have evolved adaptations to deal with parasitism, and many species are involved – there may be more parasitic wasps even than beetles (p.11), for example. Parasitism is a complicated business and there are species that parasitise other parasitoids. These are **hyperparasitoids** and in the gall wasps, for example, there are several levels of such parasitism. There are many wasp and bee species that act like cuckoos – the female enters the nest of another species of bee, lays her egg in a cell and so avoids building her own tunnels or collecting food for her young – and there are variations on the theme, just as there are in birds.

Does this make the cuckoo wasp/bee lazy or sneaky? Think of all the animal life in the garden. Evolution has led to animals fitting into all kinds of spaces, between plants, logs, soil particles – or into the nests or bodies of other animals. Cleptoparasitic cuckoo bees take advantage of the food supplies gathered by their host bee. But success in this is also hard work and has its own wonderful adaptations. These bees have to find the right nests at the right stage. The host bee will try to stop the cuckoo bee from taking over her cells and may be very aggressive if they meet in or near the nest. So, cuckoo wasps and bees have evolved defensive behaviours and disguises in order to enter host nests successfully, and have often evolved to be species-specific, although many will use the nests of a small range of bee species, sometimes across genera.

Sapyga quinquepunctata, *below, is a beautiful cleptoparasitic wasp frequently seen around bee hotels. It will enter a nest of* Osmia bicornis, O. leaiana, O. caerulescens *or* O. aurulenta *to penetrate a cell with its sting and lay an egg on or near the egg of the host bee. The larval wasp consumes the egg and then the food supply left by the mother bee.*

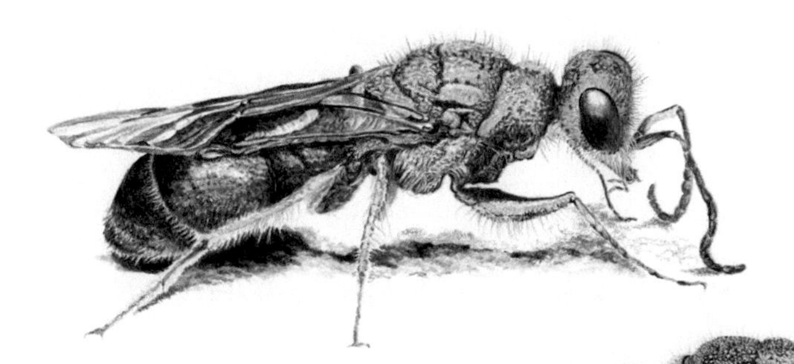

Cuckoo wasps:
Left: Chrysura radians *(host bees:* Osmia leaiana, O. caerulescens, Hoplitis adunca*).*
Below: Trichrysis cyanea *(host bees:* Hylaeus, Heriades truncorum, Chelostoma *and solitary wasps).*

Cuckoo wasps (Chrysididae; also called jewel wasps) are small, iridescent wasps of great beauty. They are difficult to spot, though, because they are restless and very fast-moving. There are more than 2,500 species globally of which between 33 and 37 are found in Britain (their classification is problematic). Their bodies are modified for their way of life – the abdomen has only three visible segments with hardened cuticles, the remainder forming a genital tube that is usually retracted. The tube is an ovipositor in the female and a poorly equipped sting. The sternites are concave. There is another concavity above the base of the antennae (which are mounted only just above the mandibles). These are defensive adaptations – when attacked by the host wasp or bee, the female curls into a ball, antennal scapes tucked into the facial concavity, head and antennae tucked tightly into the curled abdomen.

Philanthus triangulum *(the bee-wolf, right) is a wasp that paralyses honeybees then carries them to its nest as food for its larvae.*

Cerceris rybyensis *(below), hunts several mining bee species but usually predates only one species per nest (it will have more than one nest) and on females carrying pollen. Between five and eight bees per cell are used.*

Cuckoo bee cleptoparasites have evolved as many as 20 times in several bee families, often using closely related bees as hosts. There are six cuckoo bumblebees in Britain (in the subgenus *Psithyrus*), which are now classified in the genus *Bombus*, with their hosts. There are cuckoo bee species in the Halictidae (e.g. *Sphecodes*, p.93), in the Megachilidae (*Coelioxys*, opposite, and *Stelis*) and in the Apidae (*Nomada* p.109, *Epeolus* and *Melecta*, below). A detailed study of cleptoparasitism in the Apidae has shown that the trait is an ancient one – evolving in the Late Cretaceous period, 95 MYA, and then three further times, 23, 21 and 19 MYA respectively – compared with around nine times in the Halictidae. There are many cuckoo bee species – around 2,500 globally including perhaps 850 within *Nomada* alone. In 2018, 15 new species of *Epeolus* were discovered by re-examining North American museum collections. The females, of course, have no special pollen-collecting adaptations.

Melecta albifrons, above, is a striking bee, with white patches of hair that give the abdomen and legs a frilled appearance, a 'collar' of long pale or tawny hairs on the thorax, and hairs on the face that emphasise the projecting clypeus above the mouthparts. This combination, along with its size makes the species distinctive when hanging around populations of its host – our hairy-footed flower bee (see p.45). You will only see one or two, however, at any site. This makes them difficult to study: there is little information on breeding behaviour. *Melecta* can be seen entering nests when the *Anthophora* female is absent, and conflict may ensue if she returns. In related species the completed nest of the host is penetrated by the female and the egg laid on the provisions in the cell. The cell wall and nest hole entrance may then be repaired to disguise her work. On hatching, the first larval instar has an armoured head and long curved mandibles used to destroy the host's egg – and perhaps any other *Melecta* eggs or larvae that are in the cell. These mandibles are shed at the first moult. The fully-grown larva spins a cocoon and pupates in late summer, becoming an adult in early autumn and overwintering in the cell in this form. There are 59 species globally but only one or two in the British Isles – the second species, *M. luctuosa*, is probably extinct.

Related to the *Megachile* bees, *Coelioxys* is another cuckoo genus whose females are quite distinctive (below). The abdomen is long, thin and conically shaped and has a sharply tapered tip. The sixth sternite in most species projects beyond the tip of the sixth tergite (below right), emphasising the 'sharp-tailed bee' effect. The male's abdomen is not so tapered, but still looks a little unusual when seen in the garden (above). A close look reveals pairs of distinctive spines on the sixth tergite (above centre). The genus is also unusual in having hairs on the compound eyes – the only bee to share this feature with honeybees (p.38). The hairs on the sides of the thorax and legs of two species are unusual, flat and scale-like, totally covering the surface. The sharp tail is seemingly used for penetrating the seal of the nest of a *Megachile* host – the egg being laid into the cell through a slit. Some females may also enter nests before completion. As with *Melecta*, the *Coelioxys* larva has sturdy curved mouthparts – short and sharp in the first instar, long and sickle-shaped in the second, and shed after that. The seven British species are quite difficult to separate without sample specimens – these images show either *C. inermis* or *C. elongata*.

The nomad bees (*Nomada*) are a large genus – 36 species in the British Isles including two added since 2016. Many species look very wasp-like, being sparsely haired and often with striped, yellow-and-black abdomens. They can be difficult to spot because they are small and often move continuously and erratically whilst searching for host nests. *Nomada* chiefly parasitise *Andrena* mining bees; the best chance of finding them is around *Andrena* aggregations in the spring when more than one nomad species may be found at the same time. They will also use *Lasioglossum, Melitta* and *Eucera* species. Host nests are found using scent (see opposite) and the eggs of *Nomada* are often inserted into the cell wall of the host nest during construction. More than one egg may be laid in a host nest by a *Nomada* (and more than one female may enter a host nest). On hatching, the larva is very mobile and, as with *Melecta* and *Coelioxys*, has long sharp mandibles to find and destroy other *Nomada* eggs and those of its host (which it will also consume). It then grows by eating the provisions left by the host.

The number of *Nomada* species and their similarities make identification challenging. The bee pictured above could be a male of either *Nomada goodeniana* or *N. marshamella* (the markings on the abdomen being unclear). Both are common species over much of southern and eastern Britain and southern and eastern Ireland. The host species of *N. goodeniana* include *Andrena nitida* and *A. nigroaenea*: both occur in the same garden as the *Nomada* above. The primary host species of *N. marshamella* is *A. scotica*, not recorded in this garden, though another possible host, *A. haemorrhoa*, is common. The photographs below probably show a male (left) and female (right) *N. flava*. They were taken close to a small, active aggregation of *A. fulva* (p.71). *A. fulva* is considered the main host of *N. panzeri*, which is very difficult to distinguish from *N. flava*, though there are some records of *N. flava* also parasitising *A. fulva*. Take your pick!

Above, this is most likely N. goodeniana*, showing considerable agility and the flexibility bees have to hang onto the underside of flowers (*Allium triquetrum*). Although seeking nectar not pollen for a nest, you can see that this bee will still be transferring* Allium *pollen on the hairs of its face. Left, the dark rear surfaces of the antennae and the yellow on the face show this to be a male.*

Nomada males produce a pheromone from antennal glands that mimics the female pheromone of the host bee. The nomad male patrols sites likely to attract their own females. On mating, the male passes on the mimic pheromone to the female *Nomada*. She then fans this pheromone (*N. fabriciana*, right) around the nest of a host she has just penetrated to disguise her visit. Adult *N. fabriciana* vary in size, perhaps according to host species. Smaller bees hatch from, for example, *Andrena bicolor* nests (left) and larger individuals from *A. nigroaenea* nests.

The order of true flies – **Diptera** – is a very large group of insects (p.4) and of great significance to humans, as there are many species with parasitic lifestyles that exploit vertebrates – both us and our domestic animals. The 16,000 parasitoid species are responsible for a wide range of serious diseases and conditions. Bees have several dipteran predators, parasites and parasitoids, which can cause changes in adult behaviour (see p.56 for the impact on mason bee behaviour) and physiology and so impact breeding success. Large aggregations of mining or mason bees are inevitably attractive to equally large numbers of predators and parasitoids. Creating spectacular bee hotels is tempting – but if they are successful in raising concentrations of solitary bees they will be a magnet for parasites. Parasites can rapidly reduce a bee population – a perfectly natural process. They have their own beauty and fascination, and a good bee hotel is nothing without a patrolling *Sapyga* or *Gasteruption*. The appearance of *Bombylius* in early spring, hovering with incredible skill and precisely inserting its long proboscis into the flowers of honesty, is itself a noteworthy harbinger of the many beautiful *Andrena* about to emerge.

The 'satellite fly', *Leucophora* species, top, does what its name suggests – it follows close behind the host bee's every move. This was one of several individuals shadowing female *Andrena fulva* in a small aggregation (see p.71). The smaller drosophilid fly *Cacoxenus indagator* (lower image above) uses a different tactic. It is commonly seen waiting for the opportunity to slip into the nest holes of *Osmia* species (here it is just to the right of a working female *Osmia bicornis*). Both these fly species are strict cleptoparasites using the host food supply – though the bee larvae may subsequently starve.

Bee-flies, *Bombylius* (this is *B. major*), are parasitoids of several *Andrena* species. The eggs are flicked onto the ground around *Andrena* nest holes by the hovering female (sometimes it is easier to spot the fly than the nests). On hatching, the larva enters the nest to consume first the pollen store then the bee larva (see also p.72). *B. discolor*, the dotted bee-fly, may also use *Anthophora plumipes* nests.

Spiders (*Enoplognatha ovata*, left) can be effective predators of bees, even big bumblebees. The *Megachile willughbella* male, below left, has been attacked by a spider (possibly *Pisaura mirabilis*) hunting on the Trellick Bee Tower – his left foreleg is held by a strand of web. After an hour he had seizures and died. Below, the crab spider *Misumena vatia* male (and the female, inset) lies in wait for prey; the female can change body colour for camouflage. There is evidence from the US that honeybees can detect a crab spider by its odour. Jumping spiders, *Salticus scenicus* (bottom), also prowl the bee tower looking for smaller prey such as *Hylaeus*.

There are some strange evolutionary avenues in parasitism. One of those affecting bees is the parasite *Stylops*. In the order **Strepsiptera** (twisted-wing flies, though they are probably closer to beetles), the insects' lifecycles and body forms are bizarre. The female *Stylops* remains anchored within the abdominal cavity of an adult bee. She has no eyes or antennae and her body remains grub-like. Her eggs and larvae stay within her own body during growth. A tiny larval stage (the **triungulin**) emerges from her head region (the **cephalothorax**) and when the host bee lands on a flower the triungulin leaves and waits for a new, unparasitised bee. Hitching a ride into the new host's nest the *Stylops* larva penetrates a bee larva within a cell. The bee larva then grows to maturity with the maturing *Stylops* inside. Once the bee has emerged from its nest the *Stylops* larvae pupate. The female *Stylops* remains in the bee, with only her cephalothorax protruding between the bee's tergites. The male (right, top) emerges from his puparium between the tergites of the host bee, leaves the host and flies out on his strange wings to find a new female in the abdomen of another host bee. He mates with the female by mounting the bee and injecting sperm into the female *Stylops* through the same opening in her cephalothorax from which the larvae emerge. The male has unusual antler-like antennae, an odd form of compound eye with a limited number of 'eyelets', sometimes compared with fossil trilobite eyes, and forewings reduced to stabilising rods. He only lives for a few hours. The host bee remains alive, but in many species does not develop to sexual maturity and will never reproduce. The majority of parasitised bees seem to be females, and in species that are **protandrous** (in which males emerge significantly earlier than females) parasitised females often have poor development of female characteristics (such as pollen scopae) and emerge earlier in the season than unparasitised females. This appears to be manipulation by the parasite to tie the bee lifecycle more closely to the needs of the *Stylops*: the host bee is '**stylopised**'.

A triungulin active larval stage is also found in the beetle family **Meloidae**, the oil beetles. There are five UK species in the genus *Meloe* (three others are probably extinct), one species in the genus *Lytta* (all using *Andrena* species as hosts) and *Sitaris muralis*. *Sitaris muralis* is a cleptoparasite of *Anthophora plumipes*, though it may also use the much rarer *A. retusa*. Its triungulin larvae emerge in autumn from eggs laid close to a host nesting site but remain concealed and dormant until the following spring. Then, after climbing a flower stem, they wait for a visiting female bee to transport them to a nest under construction. If attached to a male bee, they transfer to a female during mating. In the nest hole they stay on the female bee until an egg is laid then drop off and, sealed into the cell, consume the egg and its provisions. They take most of the summer to mature then overwinter as a prepupa before final pupation and metamorphosis in the next spring. They emerge as a new adult in early summer. Adults do not feed, mate shortly after emergence then seek a host nest site for new egg laying. The species is very rare in the UK. Another beetle species, *Stenoria analis*, is expected to colonise following its host the ivy bee. If it does, it may well slow down the ivy bee's spread.

The small hive beetle *Aethina tumida*, an African species, is a serious pest in honeybee hives, destroying eggs, larvae and stores. It appeared in Portugal in 2004 and Italy in 2014 and is a serious worry to beekeepers. Imports of bees are banned from Italy to prevent its arrival in the UK.

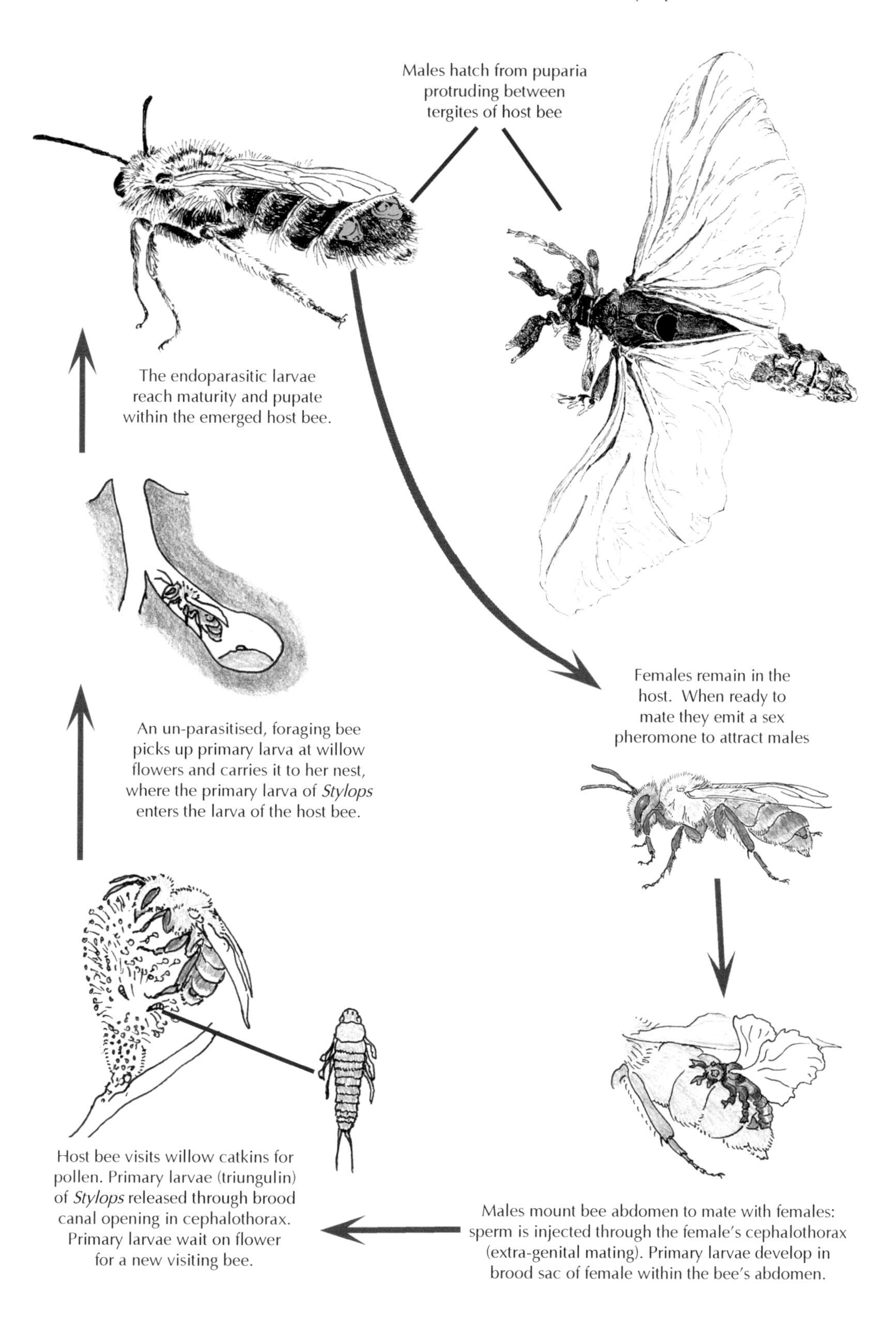
Males hatch from puparia
protruding between
tergites of host bee
The endoparasitic larvae
reach maturity and pupate
within the emerged host bee.
Females remain in the
host. When ready to
mate they emit a sex
pheromone to attract males
An un-parasitised, foraging bee
picks up primary larva at willow
flowers and carries it to her nest,
where the primary larva of *Stylops*
enters the larva of the host bee.
Host bee visits willow catkins for
pollen. Primary larvae (triungulin)
of *Stylops* released through brood
canal opening in cephalothorax.
Primary larvae wait on flower
for a new visiting bee.
Males mount bee abdomen to mate with females:
sperm is injected through the female's cephalothorax
(extra-genital mating). Primary larvae develop in
brood sac of female within the bee's abdomen.

Big changes have taken place in British wildlife over the last 50 years and are well documented in much nature writing. You can see change in action in your neighbourhood: the decline of swifts, the collapse of sparrow colonies and the absence of butterflies, as well as the arrival of 'exotics'. These changes are due to a multitude of reasons: toxic pollution, loss of key habitats for breeding or winter survival, development, persecution, even the popularity of domestic cats, and climate change. In addition, the urban heat-island effect means that impacts from wider climate change are somewhat masked in cities: the blackbirds in central London start singing in spring well before their suburban counterparts and the hairy-footed flower bee flies in February. The appearance of new species moving to Britain from the continent or into northern areas from southern strongholds can only happen if conditions are suitable, made so, perhaps, by habitat improvements (the popularity of wildlife gardening for instance) or by climate change.

Climate change – earlier springs, milder autumns and shifts in rainfall patterns – can cause seasonal behaviour shifts in our flora and fauna (the subject of **phenology**) or changes in range –northern or southern limits – or the filling of distribution gaps as a species becomes more common. A thriving population that is embedding a species' presence is almost certainly a prerequisite for range expansion. Such changes can be looked for by phenologists but they will need reliable and long-term datasets. Tracking distribution change requires good observational data, which depends on a wide distribution of knowledgeable observers. Increased media attention for phenology and the citizen science projects run by conservation organisations has attracted more observers in recent years, and social media and smart phone apps have helped NGOs improve record-keeping and data submission. Popular campaigns could produce unreliable data but confirmation and verification procedures used by environmental record centres and the National Biodiversity Network (NBN) ensure that doubtful records are flagged up. Many bee records come through BWARS.

Tracking the move north of wildlife is easier with species that have recently arrived in the UK and which are subsequently occupying more and more territory. BWARS has been monitoring the spread of the tree bumblebee *Bombus hypnorum* and ivy bee and has run recording projects for *Anthophora plumipes, Andrena fulva, A. cineraria, A. vaga, Osmia bicornis* and *Anthidium manicatum*. The maps compiled here have been redrawn from the NBN Atlas website, from the National Biodiversity Data Centre, Ireland, and other data (see notes, pp.174-179). They highlight a selection of dates to illustrate changes in range.

The first record of the ivy bee *Colletes hederae*, opposite, was in Dorset in 2001. There were already large numbers of nests and the species had been seen elsewhere in Purbeck previously. It spread rapidly along the south coast and within five years was in north Somerset. Within another five years it had reached South Wales, Cornwall and inland to Oxfordshire, then spread northeast to Flamborough and Saltburn-by-the-Sea and northwest to Wirral, to Lancashire and to the western tip of Wales. By 2021 it was found east of Edinburgh and had crossed the Irish Sea to the Raven Nature Reserve in Co. Wexford (more than 50 bees seen). It established well in Wexford and expanded its Irish range north to Wicklow in 2022 and west into Co. Carlow in 2023. The ivy bee seems unlikely to stop!

2:	29/09/02	Corfe Common
3:	29/09/03	Prawle Point
4:	24/09/04	Hastings & Lyme Regis
5:	15/10/06	Weston-Super-Mare
6:	2010 on	Broadstairs
7:	9/09/11	Tregantle, nr. Plymouth
8:	30/09/11	Oxwich, Gower
9:	01/10/11	Holywell, nr. Newquay
10:	02/10/11	Frilford nr. Tubney, Oxon
11:	20/10/11	Gribbin Head
12:	29/09/12	Lamorna
13:	18/10/13	Peckham
14:	21/08/13	Morston Marsh
15:	27/09/15	Cound, nr. Shrewsbury
16:	08/10/15	Little Orme Caernarvonshire & nr. St. David's, Pembrokeshire
17:	17/09/16	The Garrison, St. Mary's
18:	30/09/16	Flamborough Head
19:	05/10/16	Saltburn-by-the-Sea
20:	08/10/16	Heysham
21:	14/10/16	Wirral C. Park
22:	17/09/17	Abersoch
23:	16/10/17	York
24:	15/10/18	Grange-over-Sands
25:	16/09/19	Whitburn
26:	21/09/20	Skinburness
27:	26/09/20	Wetheral
28:	26/09/21	Warkworth
29:	17/09/21	Thorntonloch, nr. Dunbar
30:	12/10/21	Raven Nature Reserve, Wexford

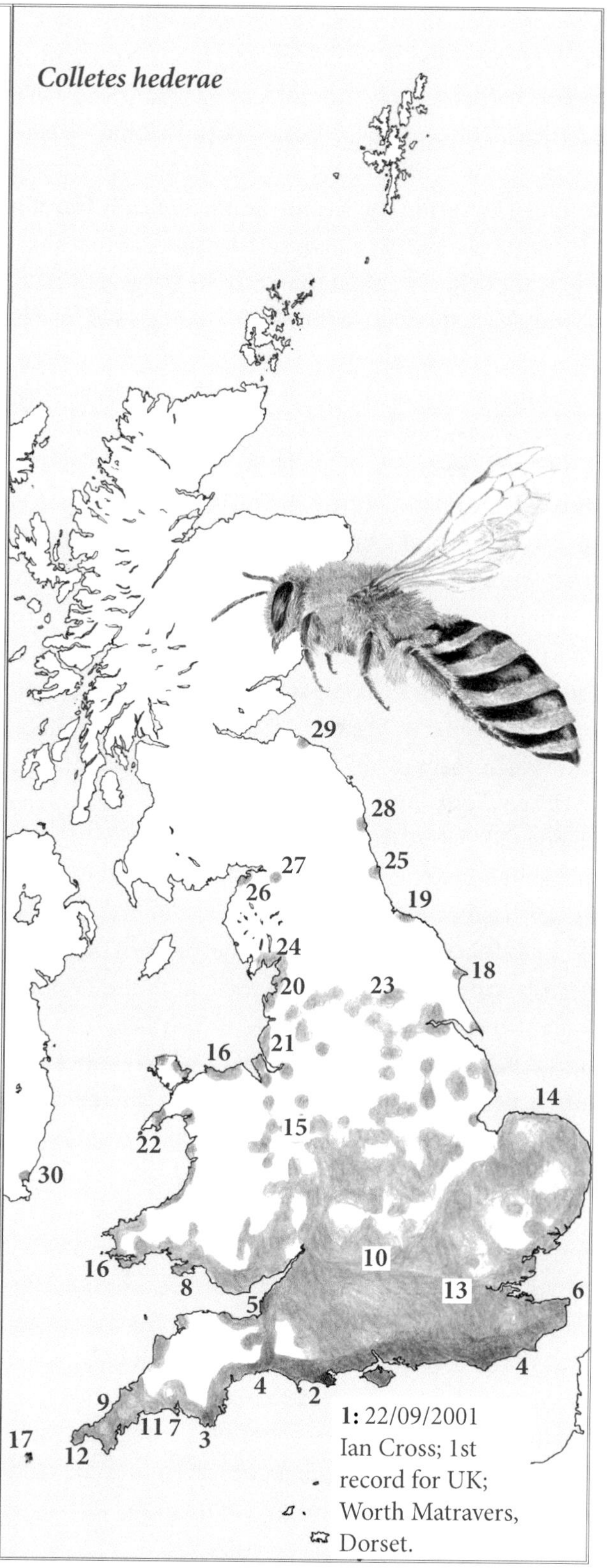
Colletes hederae
29
28
27
25
26
19
24
18
20
23
21
16
14
15
22
30
10
16
13
6
8
5
4
4
2
9
11
7
3
17
12
1: 22/09/2001 Ian Cross; 1st record for UK; Worth Matravers, Dorset.

The spread of *Colletes hederae* across the British Isles has coincided with the recruitment of far more recorders than two decades ago. The species undoubtedly arrived on the south coast before 2001, probably more than once, but once alerted, human observers have tracked it pretty well. It seems to have used coastal habitats to spread northwards and around Wales but it is also now consolidating in urban habitats and gardens (with practically every big, sunny clump of flowering ivy in south London reliably full of ivy bee females). Attributing its spread to climate change is complicated by the fact that this bee has moved into an environment where the primary pollen source for its larvae, ivy, is ubiquitous. Ivy is so frequent and, when flowering, so productive that despite being used by a host of pollinating insects (honeybees, flies, wasps) it remains a very available niche for a new bee.

Anthophora plumipes, however, is a bee with a historically well-established distribution. Its core area is southeast and central England and south and east Wales. Nevertheless, there is a historic record from 1947 for Carlisle, a record for west Pembrokeshire in 1979 and a record for Scotland in 2013. Before 1900 it was reported as widespread and locally abundant in Lancashire and Cheshire but there has been little further evidence of it until recently. Its distribution may well have fluctuated, and teasing out whether hairy-foot has been responding to climate change in the last decade or two may be tricky – hence the recording effort launched by BWARS. Reaching Kilmartin in Argyll, where it was seen feeding on lungwort in May 2020, is probably a good sign of expansion. Another sign of self-propelled expansion came in March 2022 with its arrival in Ireland at a community garden in Dublin. It was seen again there in spring 2023 but without signs of expansion.

Have these shifts in distribution been accompanied by consolidation or phenological changes? Since 2010, *Anthophora* has been seen in London, Berkshire, Leicestershire and Wiltshire in February, suggesting that it is emerging earlier than in the past. Bees emerging earlier need early flowering forage, of course, so it is fortunate that since the 1990s white deadnettle *Lamium album*, favourite forage of the hairy-footed flower bee, has been flowering from January in many parts of the UK; in the 1950s the plant did not develop flowers until March. It is common across the British Isles, enabling any dispersing *Anthophora* to find it. Assuming the absence of lengthy wet conditions or indeed drought in spring (reducing nectar production in plants), such early emergence should lead to a longer nesting season for spring-flying solitary bees such as *A. plumipes*. Infilling of a patchy distribution and expansion into new areas is thus more likely.

Gardeners have long sought out winter-flowering plants to cheer up the greys of January and February. Winter-flowering honeysuckle *Lonicera fragrantissima*, is a vigorous and popular scented shrub and is well used by all late-winter/early spring bees. Lungwort *Pulmonaria officinalis* is loved both by gardeners and *Anthophora*, is early to flower and though not native is naturalised over much of Britain. It is worth watching both these plants for your first hairy-foot of the year. Non-native winter-flowering plants (many of which are from China) may be responsible for the survival of most winter-flying bumblebees but with the phenological changes observed in some native flora, early flying bees need not be restricted to gardens.

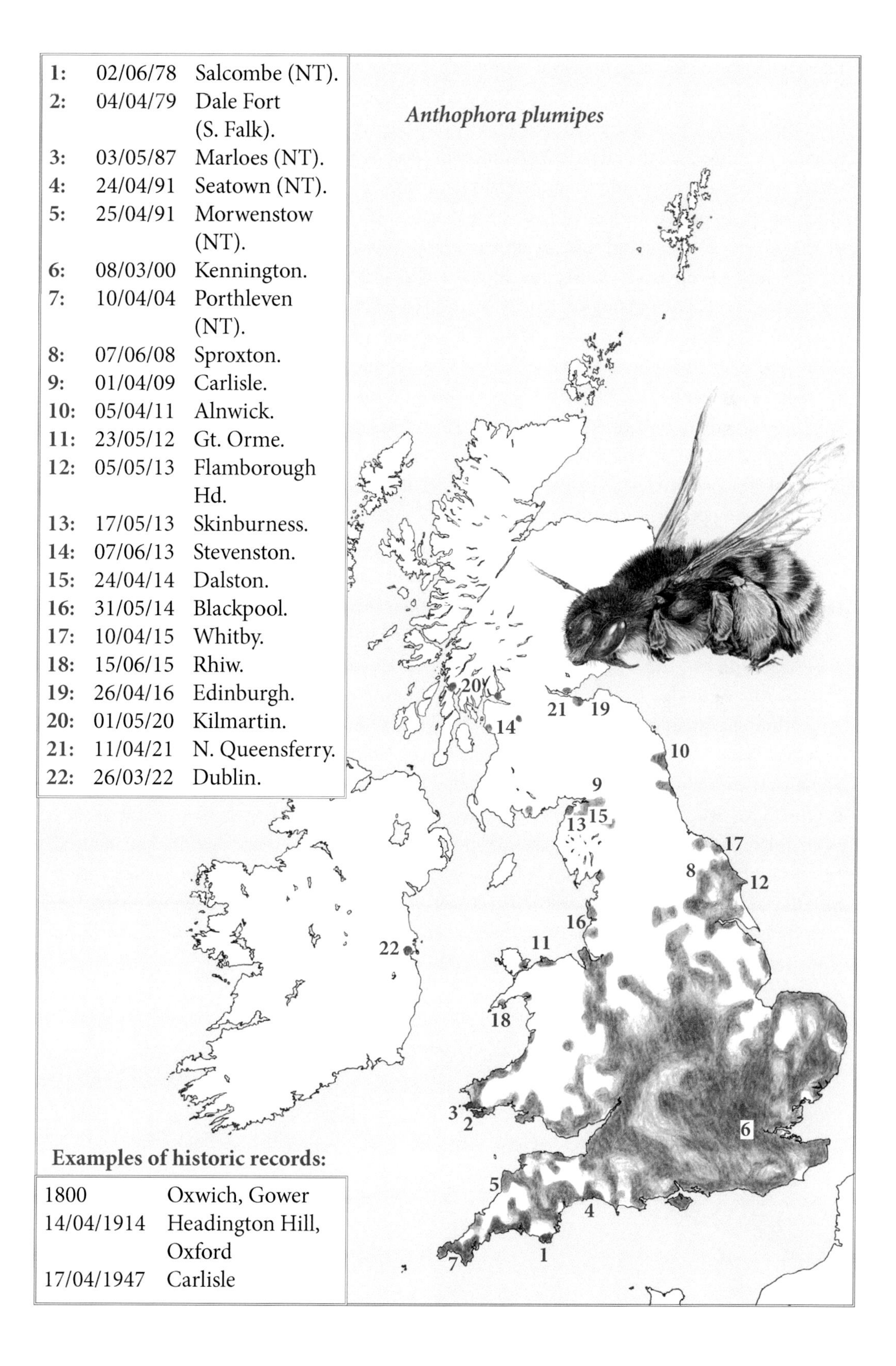

1: 02/06/78 Salcombe (NT).
2: 04/04/79 Dale Fort (S. Falk).
3: 03/05/87 Marloes (NT).
4: 24/04/91 Seatown (NT).
5: 25/04/91 Morwenstow (NT).
6: 08/03/00 Kennington.
7: 10/04/04 Porthleven (NT).
8: 07/06/08 Sproxton.
9: 01/04/09 Carlisle.
10: 05/04/11 Alnwick.
11: 23/05/12 Gt. Orme.
12: 05/05/13 Flamborough Hd.
13: 17/05/13 Skinburness.
14: 07/06/13 Stevenston.
15: 24/04/14 Dalston.
16: 31/05/14 Blackpool.
17: 10/04/15 Whitby.
18: 15/06/15 Rhiw.
19: 26/04/16 Edinburgh.
20: 01/05/20 Kilmartin.
21: 11/04/21 N. Queensferry.
22: 26/03/22 Dublin.
Anthophora plumipes
20
21
19
14
10
9
15
13
17
8
12
16
11
22
18
3
2
6
5
4
1
7
Examples of historic records:
1800 Oxwich, Gower
14/04/1914 Headington Hill, Oxford
17/04/1947 Carlisle

Like *Anthophora plumipes*, the wool carder bee has had a core distribution in the south and east of England, but the records in the NBN atlas further north and west are a little patchier by comparison. It was scarce until around 1993, when it started to increase in number and expand its range (see the late-1990's Cumbrian records opposite). It is a striking bee: both males and females draw attention to themselves with their behaviour (territorial patrolling and scraping of plants, respectively), so presence in an area will quickly be recorded. Records of the species for Scotland and Ireland probably reflect new movements. The first Irish records may have been a consequence of accidental introduction by the horticulture or timber trade via plant or compost stocks or wood (bees that nest in cavities are perhaps more prone to this than others). However, there have been populations in west Wales since the early 1990s (right) so there seems no reason why the species should not have made its own way over, as the ivy bee did in 2021 (ivy bees nest on the ground in sandy banks and therefore are far less likely to be helped by humans). The arrival point of both species in Ireland, on the southeast coast of Wexford, including good numbers of *Anthidium manicatum* at the National Biodiversity Data Centre gardens in Waterford, suggests that they made it under their own wing power. As with the hairy-footed flower bee, it has also been infilling parts of its core range, becoming more frequent in the Midlands, for example.

That the wool carder bee is perfectly capable of colonising new terrain is shown by its reach beyond its natural Eurasian distribution. It has been introduced to New Zealand and North America, where it was first recorded in 1963, and since 1996 has expanded rapidly across Canada and the US, where it is now widespread and considered invasive (see p.74). It is also in Australia, the Canary Islands, Paraguay, Uruguay, Peru, Surinam, Brazil and Argentina and can be considered the most widely distributed *Anthidium* species in the world.

Great Britain and Ireland have only one species of *Anthidium*; there are 17 on the continent. Some of these may well arrive over the Channel. A single *Anthidium septemspinosum*, the seven-spined wool carder bee, was recorded on 19 August 2021 in Hereford but as this is an inland location the bee's presence may suggest assisted arrival rather than natural colonisation. A flying male is shown below left, facing up to a resident male *A. manicatum nigrithorax*, below right. The yellow markings of *A. septemspinosum* are more extensive, including on the thorax, but markings are variable. The continental form *A. m. manicatum*, shown in photographs taken in London on p.80, also has more yellow.

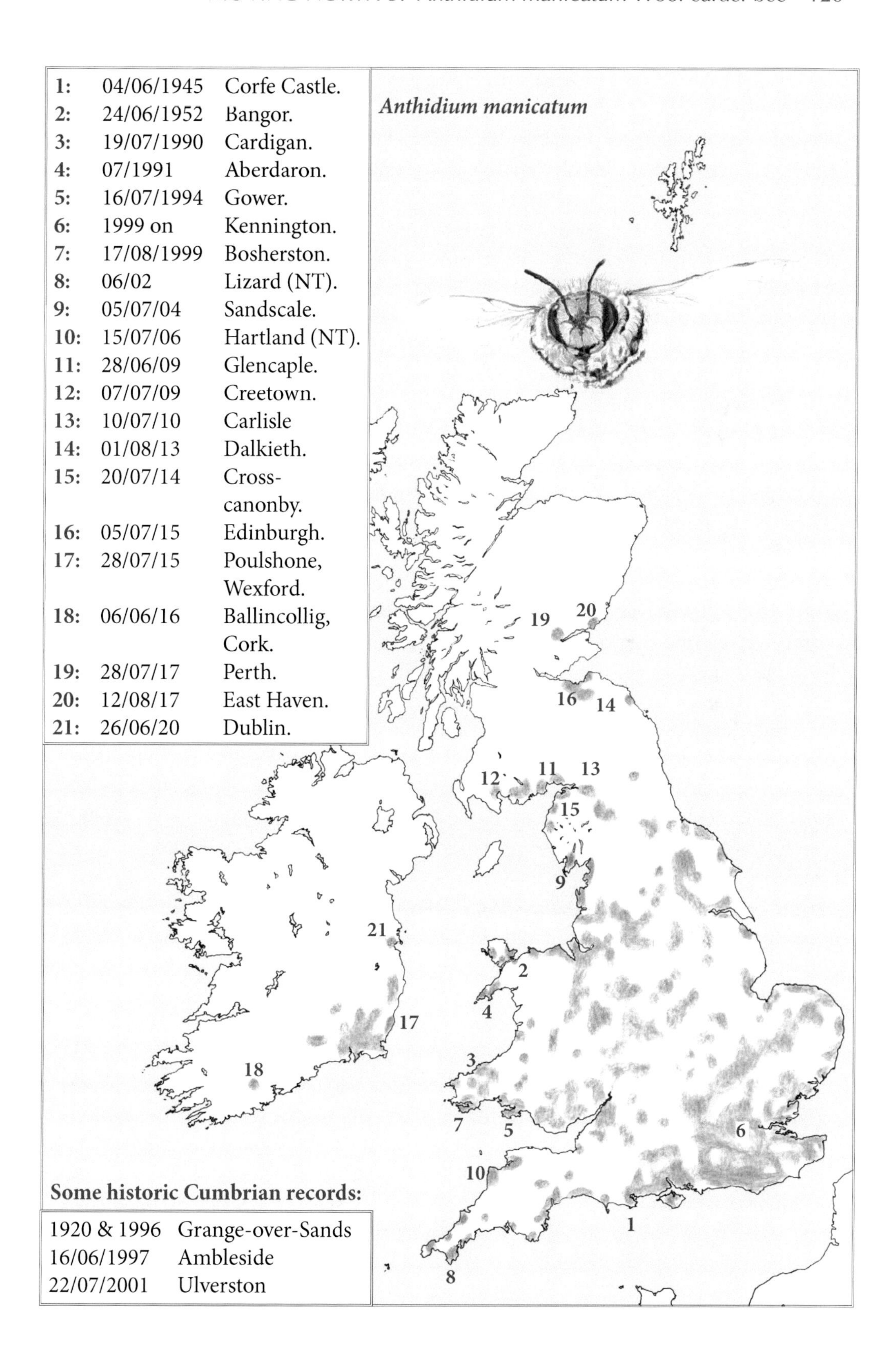
1: 04/06/1945 Corfe Castle.
2: 24/06/1952 Bangor.
3: 19/07/1990 Cardigan.
4: 07/1991 Aberdaron.
5: 16/07/1994 Gower.
6: 1999 on Kennington.
7: 17/08/1999 Bosherston.
8: 06/02 Lizard (NT).
9: 05/07/04 Sandscale.
10: 15/07/06 Hartland (NT).
11: 28/06/09 Glencaple.
12: 07/07/09 Creetown.
13: 10/07/10 Carlisle
14: 01/08/13 Dalkieth.
15: 20/07/14 Cross-canonby.
16: 05/07/15 Edinburgh.
17: 28/07/15 Poulshone, Wexford.
18: 06/06/16 Ballincollig, Cork.
19: 28/07/17 Perth.
20: 12/08/17 East Haven.
21: 26/06/20 Dublin.
Anthidium manicatum
Some historic Cumbrian records:
1920 & 1996 Grange-over-Sands
16/06/1997 Ambleside
22/07/2001 Ulverston

The appearance of a violet carpenter bee *Xylocopa violacea*, in your garden will take you aback: the species is large, dramatically dark, including the wings, and produces an unfamiliar sound. Its spectacular form probably means that the recent increase in observations is not wholly explained by more knowledgeable and aware members of the public: climate change may partially explain the proliferation of records – though there was a confirmed breeding of the species in a fence post in King's Langley, Hertfordshire in 1920. This bee nests in dead but not necessarily rotten wood – it can make its own hole – and on the continent it is well used to exploiting human structures such as building timbers or garden features. It overwinters as a robust adult in old nest holes, hollow trees or other niches. Consequently, it is possible for its nests and/or overwintering adults to be transported via the timber or horticultural trades.

The distribution of the British and Irish records is interesting, and the species has been discussed at length on the BWARS website, in the magazine *British Wildlife* and in the *Handbook of the Bees of the British Isles*. The map opposite includes as many of these records as possible and shows a very dispersed pattern. Although there is a cluster in the southeast and around London, there are also records as far north as Fife and as far west as Kinsale in Co. Cork. The almost certain breeding in an old apple tree in a garden in Shepshed, north of Leicester, includes five bees flying in February 2006 (these must have overwintered at the site) with at least one male. Several females nested later that year. Shepshed is well inland – just as *Anthidium septemspinosum* was in Hereford – but there are lorry parks and an airport nearby dealing with large volumes of freight. In 2006 there was a nest in an apple tree in Tonbridge, Kent and a February record for Borth in West Wales. If the latter bred, the first record for Ireland just across the water in Waterford in 2007 may be linked. My first sighting was a sudden glimpse in the garden at Roots and Shoots in July 2005. I then had very good views of a female in May 2006; there is also a record for nearby Peckham in September 2006. *Xylocopa violacea* may well be used to hanging around human timber works but it is also a strong flyer. There are growing populations in Normandy and on the Cherbourg Peninsula in France, increasing numbers were recorded in northern Germany in the early 2000s, and there has been a confirmed breeding in the Channel Islands. Violet carpenter bees on the south coast of England, therefore, are just as likely to be self-propelled as transported with a load of palettes. In this scattered distribution of sightings we are undoubtedly seeing a mixture of travel options for the species. This may well continue. If climate change has a role, it is by providing the bee with more amenable overwintering opportunities in the British islands. We can fairly confidently expect more confirmed breeding records in the next decade.

Another new resident breeder that also probably arrived with assistance, *Osmia cornuta,* the orchard mason bee, is a handsome relative of the red mason bee *O. bicornis*, but the male *O. cornuta* more closely resembles the female than is the case for *O. bicornis*: she has a handsome black head and thorax and rich chestnut hairs on the abdomen; males are often as large as females, have noticeably long antennae and white hairs on the face. The species was first spotted near Blackheath in London in 2014 but was initially mistaken in the photograph for *O. bicolor*; it was confirmed and accepted onto the British list in 2017, when it was also found breeding at London Wetlands Centre in west London.

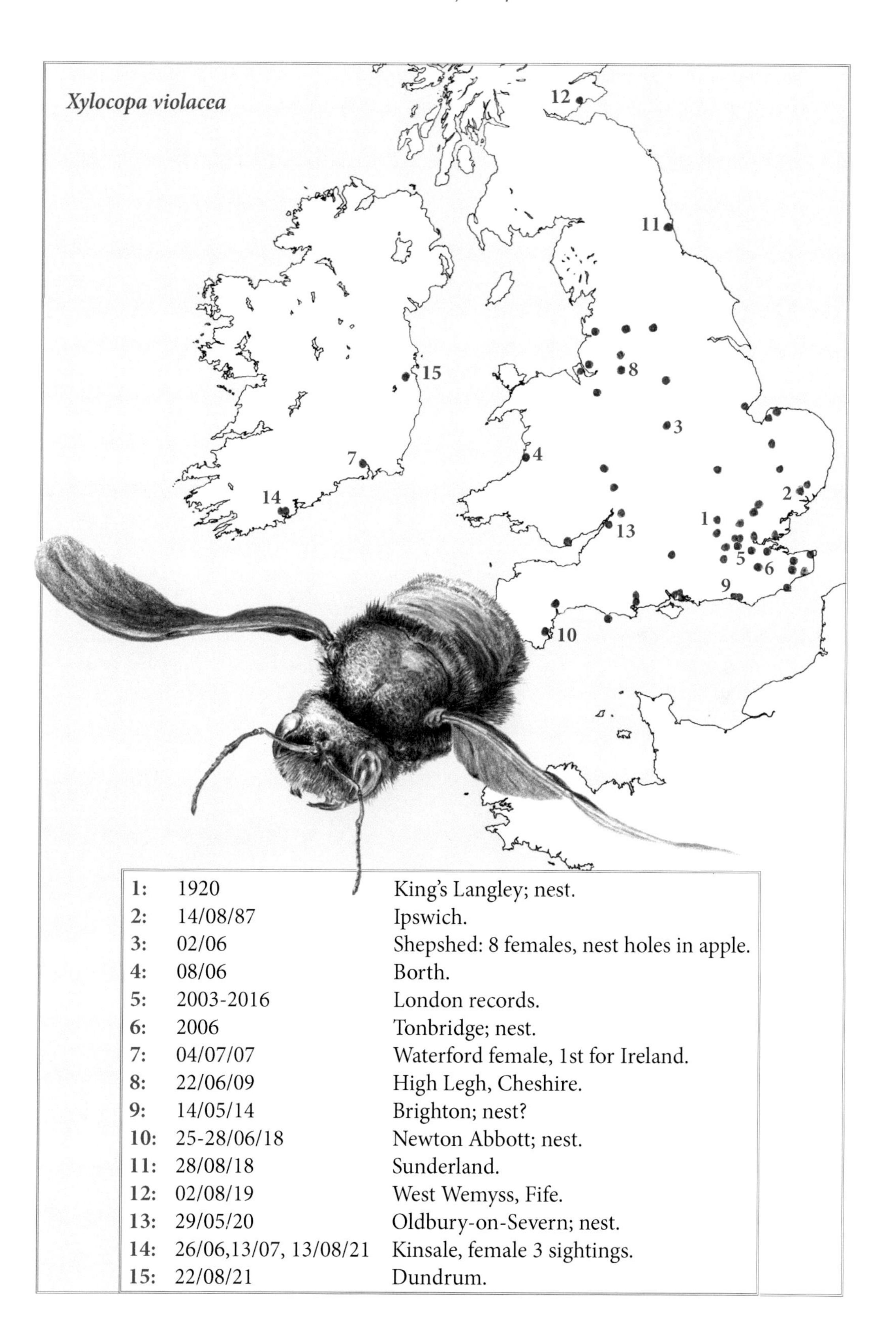

1:	1920	King's Langley; nest.
2:	14/08/87	Ipswich.
3:	02/06	Shepshed: 8 females, nest holes in apple.
4:	08/06	Borth.
5:	2003-2016	London records.
6:	2006	Tonbridge; nest.
7:	04/07/07	Waterford female, 1st for Ireland.
8:	22/06/09	High Legh, Cheshire.
9:	14/05/14	Brighton; nest?
10:	25-28/06/18	Newton Abbott; nest.
11:	28/08/18	Sunderland.
12:	02/08/19	West Wemyss, Fife.
13:	29/05/20	Oldbury-on-Severn; nest.
14:	26/06,13/07, 13/08/21	Kinsale, female 3 sightings.
15:	22/08/21	Dundrum.

Osmia cornuta is a common bee on the continent and has been used in orchard pollination for many years in Italy and Spain. It is an early flyer and is used to pollinate pears. It may have been introduced, accidentally or deliberately, even by people with second homes on the continent who brought it back in bee hotels. This is probably how the red mason bee *Osmia bicornis* reached Ireland, where the first records for red mason bee were from Dublin, Belfast (both 2003) and Cork (2009). Both *O. cornuta* and *O. bicornis* can be bought online in ready-made bee hotels, including from continental suppliers. Popular interest in bees will undoubtedly lead to more such introductions, wise or not. The orchard mason bee has been breeding regularly in Blackheath and other parts of London since 2014 and has moved quite rapidly across southern and eastern England. It reached Cheshire in 2021. *Osmia spinulosa*, which uses empty snail shells as nest sites (it is a species of open grassland and not really a garden bee), also seems to be moving north and west.

The large-headed resin bee *Heriades truncorum* (pp.95-96) is showing signs of expanding its range. Until relatively recently this species was considered rare and was restricted to the far southeast of England, notably Surrey. The accounts of 19th-century observers and collectors are a match for the more voluminous contemporary discussions about violet carpenter bees and orchard mason bees. The large-headed resin bee may also have been a (Victorian) introduction in imported timber. It was first seen breeding at Roots and Shoots in 2008 and sightings became very frequent from 2011 onwards. In 2015 it had a very long season in the Trellick Bee Tower, starting in mid-June; one female was still building on 3 October. If it is becoming more common in the southeast, expansion may be under way: it reached the Buglife grounds in Cardiff in 2019 and was seen in Norwich in 2021. The 2007 record from Shrewsbury was an outlier but may not be so for much longer. It is not now, even if it was previously, restricted to using pine resin for nest material, so expansion is not limited by that particular factor. It is happy to use bee hotels and so will find more available nest sites as it disperses than the 'ragwort growing near old fence posts', referred to by David Baldock in *Bees of Surrey*, on which it was found in the early 20th century. Even small shifts in ecology such as this, mediated or not by climate change, can assist with expansion.

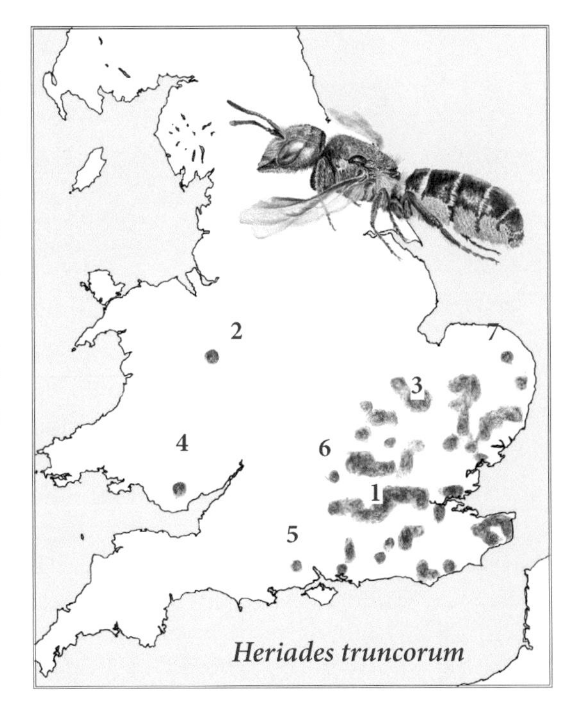

1:	06/07/1907	Chobham.
2:	25/08/07	Shrewsbury.
3:	09/07/17	Cambridge Bot. Gdn.
4:	26/06/19	Cardiff.
5:	10/07/19	New Forest.
6:	29/05/20	nr. Wallingford.
7:	21/08/21	Norwich.

Osmia cornuta*: males, left, can be as large as females and with noticeably long antennae. This male has lost his left antenna, perhaps during a competition for a female, right.*

Moving north: running out of land

Moving north is all very well if you are a southern species but for the flora and fauna of the uplands in the north or west of Britain climate change could pose challenges: moving upwards or northwards you begin to run out of space. There are some bee species with mainly northern distributions that are already under population stress. These include three species of *Osmia*, the closely related *O. inermis, O. uncinata* and *O. parietina*. *Osmia inermis* and *O. uncinata* are restricted to the Scottish Highlands. *O. inermis* is particularly rare, with perhaps only four sites in the Braemar/Blair Atholl area, all on exposed upland grassland habitat. *O. uncinata* is in a slightly better position because it uses open woodlands at lower altitudes, notably the remnants of the ancient Caledonian pine forest, areas of the Spey Valley and locations stretching north to Sutherland and Ross and Cromarty. The third species, *O. parietina*, has a more southerly distribution and is found in north Wales, north Lancashire and Cumbria, although there are also records from the Borders and some parts of the Highlands, where the bees often use limestone pavements and drystone walls. All three species have a Eurasian boreal or alpine distribution. Of mining bees, *Andrena ruficrus* is mainly Scottish and is seen particularly around Inverness; this species has also been recorded in the northern Pennines and Yorkshire. *A. tarsata*, though more widely distributed including in Ireland, is restricted to heaths and moors. It has a boreal and alpine distribution from Iceland to Scandinavia, Spain, Siberia and China and, though widespread in Britain, is rare, local and declining in the south. There is also the northern colletes *Colletes floralis*, a distinct 'West Briton': Ireland has important populations. The species, which needs coastal, dune and machair habitats (flower-rich grasslands on stabilised shell-sand), is also found in the Outer and Inner Hebrides, Ayrshire and at one site in Cumbria. Its Irish populations are of conservation concern and the subject of research; the RSPB is making great efforts to help the Hebridean populations. These populations are highly significant globally. The species flies in summer and can be particularly affected by cool, rainy or stormy conditions. For this reason, climate change is a potential danger for a species already seriously affected by habitat loss (changes in grazing and land use, particularly development for golf courses and housing). There are also mountain and northern bumblebee species attracting concern in Britain and Ireland, for example the great yellow bumblebee *B. distinguendus*, which is also a machair species.

A link between environmental conditions and sociality was introduced on pp.61-62 in the discussion of social plasticity in the orange-legged furrow bee *Halictus rubicundus*. Early Miocene (20–22MYA) eusocial evolution in the sweat bees (of the family Halictidae) appears to have coincided with periods of global warming. Considering the research on social plasticity in *Halictus rubicundus*, a paper by Schürch and colleagues in 2016 explored the future changes that might occur in populations of this bee in Britain subject to a warming climate. *H. rubicundus* is common and widespread throughout the British Isles from Cornwall to Caithness and westwards in Ireland to the Mullet Peninsula, Co. Mayo. Northern or high-altitude populations of this bee are entirely solitary, whilst those in the south and east of England and Ireland and at low altitudes are mainly eusocial. Southern bees may remain solitary in poor-weather years, but northern populations have never shown sociality even in good years. Nevertheless, when females from a solitary site in Belfast were translocated to a southern location (see p.62), 46% of them went on to produce social nests. So the obvious question is: if these socially polymorphic bees can switch from solitary to social nests (and vice versa) so easily, will climate change lead to northern populations of *H. rubicundus* becoming increasingly social?

Producing a social nest with workers has consequences for the foraging and provisioning behaviour of the foundress female (the queen): she needs to start foraging and building earlier in the spring. Foraging behaviour is also plastic – in the translocation study females altered their foraging behaviour according to their environment. Foundress females in the south would forage later in the day and less intensively than northern females, who had more temperature and daylight restrictions on their behaviour. This implies that southern females are less time-stressed, although in good weather northern females only attempt one brood per year so could be less time-stressed overall. There is thus a relationship between temperature and foundress activity and between that activity and the number of workers produced. The 2016 paper by Schürch et.al. used simulated weather data to predict future numbers of workers for scenarios with low- and high-CO2 emission conditions. Daily maximum temperatures were translated into numbers of provisioning trips using observational data, then the number of provisioning trips was used to predict numbers of workers produced by an average foundress under the different climate scenarios. Results indicated more foundress females would be active and that their activity would be more intense when temperatures reached 20–21°C; that the number of workers produced per nest would be determined by spring foundress provisioning activity; and that for all future emissions scenarios *H. rubicundus* worker numbers are projected to increase across Britain. This increase would be most evident in the east and southeast but it is most dramatic in the north and in Scotland. For the climate optimists amongst us the change in sociality of these bees under low-emissions forecasts will be less spectacular but social nests will nevertheless become common over most of Britain.

A typical nest site for a large aggregation of Halictus rubicundus. *This south-facing slip scar in the bank of a river in Northern Ireland has at least 300 nests. The panoramic shot below has most of these marked with a 'digital stake' – a small hash sign. The females here will still be solitary: April is, often enough, a difficult, if not cruel, month. A site like this will also have to cope with periodic river flooding.*

The research discussed left has been extended recently by Rebecca Boulton (Boulton and Field, 2022) who has studied the evolution of sensory and social systems using *H. rubicundus*. In social insects the antennae have greater densities of chemoreceptive hairs (p. 38) to facilitate, for example, nest-mate recognition (by detecting the hydrocarbon profiles of the cuticles of others), receiving information about nest state and larval development, and to help maintain worker caste systems where these occur. This study found that the density of olfactory plates and hairs differed between southern and northern populations, and that, after translocation from Migdale in northern Scotland to the Knepp Estate in Sussex, the young of Scottish bees developed with higher densities of olfactory plates and hairs than their Scottish-born parents, due to higher soil temperatures (2-5°C warmer) during development. The conclusion was that olfactory sensitivity is plastic in *H. rubicundus*, supporting the development of sociality by favouring olfactory social communication when environmental conditions are suitable. Interestingly, she also found that the Belfast population studied previously had higher densities of olfactory hairs than those in Scotland and equal densities to populations in southwest England. If these densities indicate a predisposition to develop sociality it could explain earlier results of translocation between Belfast and southern England. The work supports the general validity of the predicted expansion of sociality in this bee across the UK described opposite.

Another climate-mediated behaviour change is the increasing number of early spring-flying species recorded in autumn. Examples are *Andrena scotica, A. cineraria* (both foraging on ivy and collecting pollen) and *A. nitida* flying in autumn 2018; *A. bicolor, A. dorsata*, also collecting pollen, in October 2018; and *A. nitida*, seen in October and November 2020. The National Biodiversity Data Centre has a record of a red mason bee from August 2015 in Lisburn, Co. Antrim.

SPRING IN THE WILDLIFE GARDEN

Habitats: sunny walls with soft mortar for Anthophora*; mounds of debris below the wall and above the pond for* Andrena*; old gate posts drilled for* Megachile, Osmia, Heriades, Hylaeus*; old oak gate with knot holes for* Megachile *and beetle larvae holes for* Hylaeus*; old paving laid on sand for* Andrena*; wet clay at the water's edge for* Osmia bicornis*; mixed spring forage - cowslip, grape hyacinth and geranium for long-tongued* Anthophora, *dandelions for* Andrena.

The diversity of bees we have met in this book may make identification of the bee in your garden seem daunting. Many of the smaller *Andrena* mining bees are the 'little brown jobs' (p.91) of the bee world, whilst the fast-moving tiddlers in *Lasioglossum* and *Halictus* have disappeared before you can even focus. I have tried to highlight species that you might find easily in a decent urban garden and that you will find easier to identify. It is worth being realistic to begin with about what you can achieve without more specialist training, but it is also worth having a go. Until relatively recently there was little ID advice for amateur naturalists. Web-based information is now prolific but I suggest using caution with some sites. Steven Falk's Flickr site (Collection: Apoidea) is the best place to go for help with bee ID, together with his *Field Guide to the Bees of Great Britain and Ireland*. The *Handbook of the Bees of the British Isles* is the most comprehensive manual for identification with detailed dichotomous keys (you pass through each stage of the key after a 'one or the other' question about your sample until you reach a single ID point), but it is large and expensive and only for when you get really involved!

Start by developing your observational skills: use your peripheral vision to pick up unusual movements and strange flight paths, such as the sudden fly-past of a flower bee or wool carder bee, and be alert but without frantically attempting to follow every tiny zig or zag in a fleeing speck. Be patient and keep watching – that fleeing speck could well return to the foraging area you have just disturbed. Train yourself to pick up details in the garden – that tiny patch of orange on a green leaf could be a *Nomada* cuckoo bee. Also be aware of the season: the fast flight spotted in June won't be of a hairy-footed flower bee but could be one of its relatives such as *Anthophora quadrimaculata*. If you get close enough, does it have a pollen scopa under the abdomen (family Megachilidae) or on the legs (Andrenidae, *Anthophora* and other mining bees)? Or perhaps there is no pollen scopa at all, which could make it a cuckoo bee, a yellow-faced bee *Hylaeus* or a male bee. Remember that males and females of the same species are often quite different. This might seem to add even more ID difficulty but, with some exceptions, males are slimmer, smaller, lack hairy scopae, often have pale hairs on their faces and have noticeably long antennae. A male is likely to be looking for a female and using one of a few tactics to do so. So look out for more of the same – the males of many species will gather where females are likely to be. Or it may be one of the more territorial species that will return to the same area after foraging, or pose conveniently for photographs when resting after its fevered searching flights. Females may give you other clues to their identity, such as the nature of their nest site (on flat ground or in a bank, in a particular type of substrate, alone or in aggregations) or the materials they are bringing in (mud/masticated leaves/resin). Association is also useful – a cuckoo bee species can lead you to the presence of its previously unseen host – or vice versa.

However observant you are, and however quick to respond to brief and sudden movements, you will need more assistance to get beyond broad identification to family or genus. For this you will have to capture the bee somehow – physically or with a camera. Both need skill. Physical capture does not have to mean damage or death to the bee but you will need confidence to avoid harm. There is good advice on techniques in Steven Falk's field guide, and watching an expert will convince you it is possible to master the art. You will need a large, soft entomologist's net on an adjustable handle (shorter will be better in most gardens). Watch your bee's behaviour carefully, anticipate its movements as far as possible then use a decisive sweep of the net, give a rotating flick to the handle and form a fold in the net closed by its rim with, hopefully, the bee(s) trapped inside the fold. Steven Falk (below left with a long-handled net in a forest of alexander plants) then recommends raising the net to eye level facing the sun before inserting a bug pot (assuming you don't have his dexterity to trap bee legs harmlessly between forefinger and thumb – below right, not to be attempted with a female big enough to sting!). You could also quickly and gently drop the net to the ground to keep the bee trapped, allow the bee to settle then work your way in with a bug pot. Holding the net up with one hand will encourage a restless bee to fly to the top. With the male *Anthophora quadrimaculata* at Roots and Shoots, I wanted to be sure that I didn't also have *A. bimaculata* in the gardens. He was moving far too fast to get a good look at his face (to spot the two black marks below the antennae, opposite right), so I quickly netted him, transferred him carefully into a petri dish to restrict him and enable photography, then released him again to continue, unfazed, patrolling the lavender (as on p.97).

Stalking with a camera is good discipline: you spend slow time with bees. You will see detail on macro images that your eye could never pick up. A digital SLR camera with a decent macro lens can be expensive, bulky and heavy. Many bridge cameras will focus to zero cm from the lens, though few will produce an image in which the subject fills the frame when held this close. The *Handbook of the Bees of the British Isles* offers more advice in essays by Paul Brock and Jeremy Early, and the website **insectsandflight.com** has advice on techniques and equipment. Good images come with a combination of patience and skill. Look out for bees resting – especially in spring when cool air temperatures mean they will spend time sunbathing. *Anthophora* are more difficult in this respect as they seem to rest so little (especially the females) and rarely settle for any time on a flower. Stalk carefully and smoothly, avoid over-shadowing and take successive pictures with the bee at different angles as you approach for maximum information. Move ahead of foraging bees and focus on a likely flower, then hold the focus with the shutter button until the bee moves onto the flower. At bee hotels a tripod can help photograph approaching bees. Most digital SLR or bridge cameras have good video these days. Fast-moving *Anthidium* males, for example, are a big challenge – a good test of your skill and patience!

Your identification starting points will be Steven Falk's Flickr site (http://tinyurl.com/nrywslu), his field guide and the BWARS website (www.bwars.com). The latter has an excellent gallery, identification guides, advice for beginners and biological information on many species. BWARS is a membership organisation, runs ID courses each year and has a downloadable guide to bees in Britain. With a good selection of images you can try using the keys in Falk's guide. Do this several times, though, being cautious and exploring different possibilities, until you come to a best possible identification, which is often not conclusive. Photographs, even several of one individual, will not provide sufficient information to follow a key properly – the keys are designed for a mounted specimen. If you are confident of your ID you could register your records on one of the BWARS recording projects. Their regular projects focus on the more easily identified bees. For Ireland the All-Ireland Pollinator Plan (https://pollinators.ie) has a Solitary Bee Monitoring Scheme to which you can submit records. NatureScot has produced a guide to Scotland's wild bees. Hymettus (http://www.hymettus.org.uk), the aculeate research group, has information sheets on several solitary bees and reports on some of the rarer species. When you have developed your confidence, photography and ID skills you might also register with iRecord, the online register for biodiversity in the UK run by the Biological Records Centre (BRC). Records uploaded to iRecord are validated before they are accepted into the NBN. If you have already sent information to BWARS, a local environmental records centre or the Irish National Biodiversity Data Centre, it is not necessary to submit them to the BRC or NBN again.

Anthophora plumipes female approaching Pulmonaria flowers (Photo : Robin Williams)

Many citizen science projects now operate. These include BioBlitz events run by The Natural History Consortium (along with other projects) and the UK Centre for Ecology & Hydrology's pollinator monitoring scheme 'POMS', which is not limited to bees. It includes 'Flower–Insect timed counts' (FIT count), a standardised and repeatable counting system for particular flowers or flower groups. Friends of the Earth's Bee Cause project, which began in 2012, includes the Great British Bee Count and has downloadable resources, and between 2016 and 2019 Open Air Laboratories ran its Polli:Nation project with Learning Through Landscapes – other survey resources are still available. Bumblebee Conservation Trust's BeeWalk Programme is focussed on bumblebees but recorders often register other species.

Further Reading (more in the Notes on Illustrations, p. 151 onwards)**, books:**

Solitary Bees by Ted Benton and Nick Owens is the most significant recent book on solitary bees (Harper Collins, 2023). It is highly detailed and as comprehensive as you would expect from these authors and the Collins New Naturalist series.

The Solitary Bees: Biology, Evolution, Conservation by B.N. Danforth, R.L. Minckley & J.L. Neff (Princeton, 2019) is a definitive text aimed at those more advanced. It highlights complex tales otherwise difficult to find in the scientific literature.

Handbook of the Bees of the British Isles by George R. Else & Mike Edwards (The Ray Society, 2018) is two hefty volumes covering every species in the British Isles (to date of publication). The first volume has dichotomous ID keys. The first set of keys gets you to the correct genus; the second to your species. There are essays on bee anatomy, keeping records and photography, collecting samples and examining the pollen collected by bees, bees and their environment, bees and flowers, bees as prey for other species, populations and conservation, and land management practices and their impacts on bees. The second volume has individual species profiles with distribution maps and descriptions, nesting biology and behaviour, pollen collected and other flowers visited, the flight period and parasites and predators for that species. The photography is excellent.

Solitary Bees by Ted Benton (Pelagic Publishing, 2017), part of the Naturalists' Handbooks series, includes extensive information on the ecology of solitary bees and has a very good reference list and short keys to get you to the correct genus.

The Bees in Your Backyard, A guide to North America's Bees by Joseph S. Wilson & Olivia Messinger Carril (Princeton University Press, 2016).

Field Guide to the Bees of Great Britain and Ireland by Steven Falk and Richard Lewington (Bloomsbury, 2015). There is a very useful introductory section and comprehensive ID pages and keys. It contains all its content in a sturdy, portable format. This guide and Falk's Flickr site offer almost all that you will need when seeking our bees.

The Bee: A Natural History by Noah Wilson-Rich (Ivy Press, 2014). Mainly focussed on honeybees but with good coverage of solitary bees and a bit on evolution.

Bees: A Natural History by Christopher O'Toole (Firefly Books, 2013). This is almost an update of his *Bees of the World* (below) with many personal anecdotes and some big photographs.

Plants for Bees by William D.J. Kirk & F. N. Howes (IBRA, 2012) is a comprehensive guide for gardening with bees in mind. You can check plants you might grow against their likely attractiveness for different groups of bees. Plants are colour-coded for importance to honeybees, short-tongued bumblebees, long-tongued bumblebees and solitary bees. Excellent essays from Norman Carreck, David Aston and Sally Bucknall on plants for honeybees, Jane Stout on plants for bumblebees, and Christopher O'Toole on plants for solitary bees.

Bees of Surrey by David Baldock (Surrey Wildlife Trust, 2008). Surrey has many species.

The Bees of the World by Charles D. Michener (2nd edition, Johns Hopkins University Press, 2007) is big and fascinating. It can be viewed online.

Bees of the World by Christopher O'Toole & Anthony Raw (Blandford, 1991) is an excellent read still, if now old (the publisher put a hoverfly on the front cover of a later edition!).

Evolution of the Insects by D. Grimaldi & M.S. Engel (Cambridge 2005) is superb if you have a particular interest in evolution.

Ashy mining bee

Andrena cineraria*: on a sunny, sandy mound above the pond:*
F*: one female starting a new nest;*
FN*: two females inside nest holes;*
M*: three males;*
N*: eight nest holes.*

Tawny mining bee

Andrena fulva*:*
F: one female foraging on dandelion;
FN: one female at nest hole;
N: five nest holes.

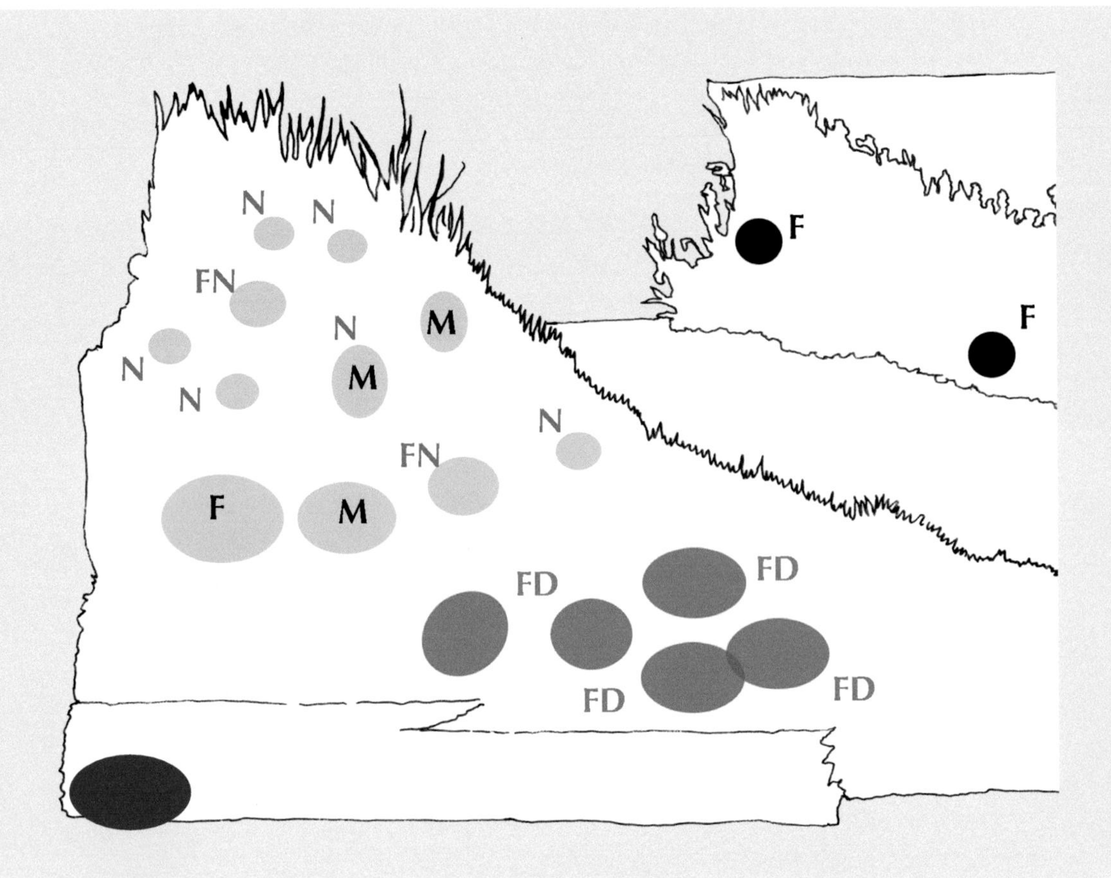

Non-bee invertebrate:

Great pond snail
Lymnaea stagnalis
feeding on
Water crowfoot
Ranunculus aquatilis

Red mason bee

Osmia bicornis*:*
FD: *Five females digging mud by the pond;*
FM*: Two females carrying mud;*
MP*: Mating pair;*
FP: *Four females carrying pollen;*
FN: *Three females working at nest holes in sunny fence post;*
MN: *Nine males hanging around, or in, holes;*
N: *54 nest holes in the right hand post and 14 in the left hand post (larger holes).*

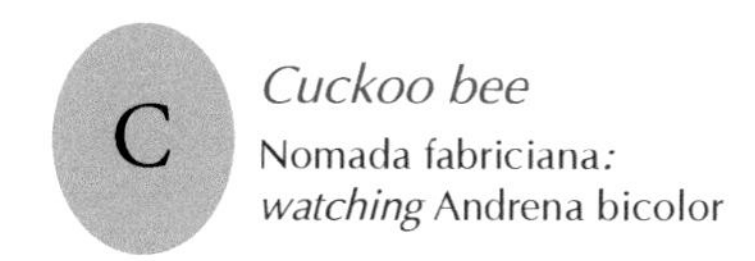

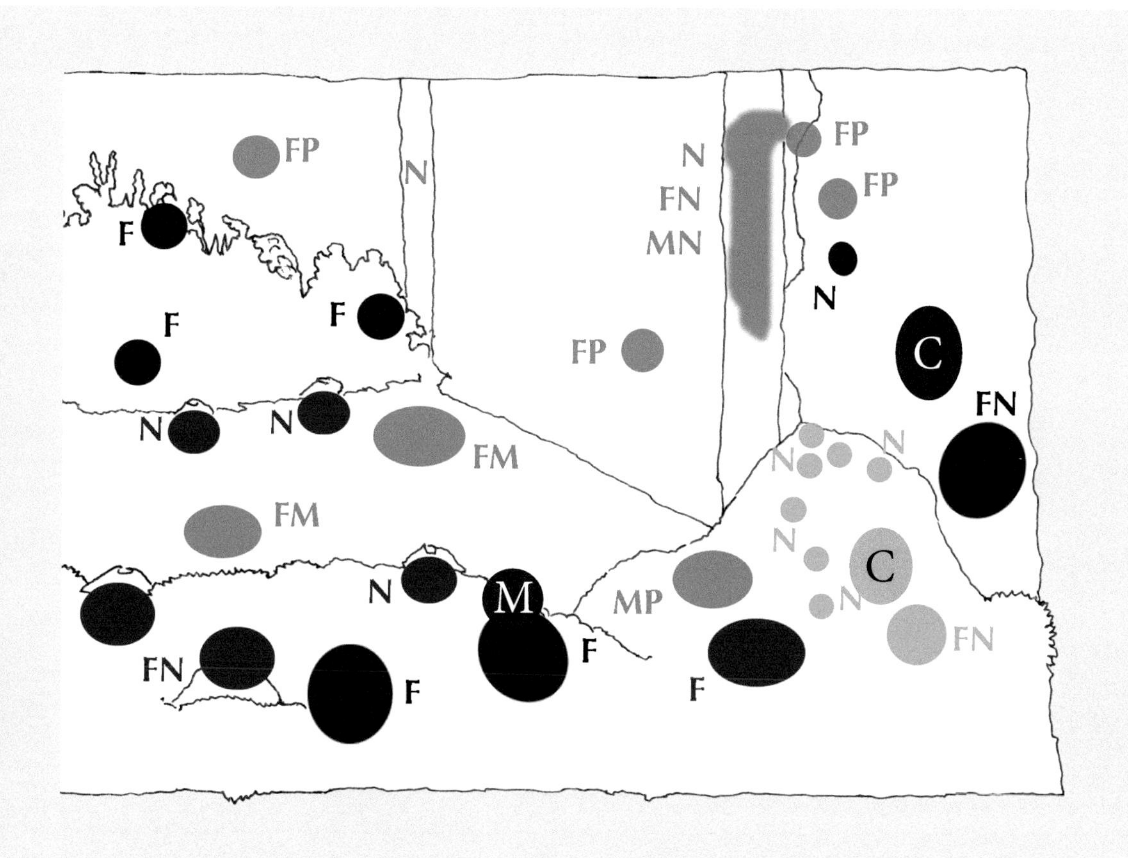

Hairy-footed flower bee

Anthophora plumipes:
F: seven females foraging on cowslip, grape hyacinth and rock cranesbill;
FN: one female nest building;

: one male trying to mate;

N: two nest holes.

Mining bee

Andrena bicolor:
FN: *one female at nest hole;*
N: *eight nest holes (and a half?), on a sandy mound by the wall.*

The Trellick Bee Tower was built in 2010, sponsored by the Community Service Volunteers' (now Volunteering Matters) Action Earth programme. It was a scaled model of the Trellick Tower in Kensal Town, London (pp.57, 77) a modernist building designed by Ernö Goldfinger. The 'lift tower' (the post-like structure on the left in the photos) originally had plastic observation tubes, in which mason bees nested and through which visitors could see them. Few nests in the tubes were successful, however, probably due to high humidity in the plastic, and after two seasons they were removed. The tower was refurbished several times with bee blocks replaced and timbers and towers added, but by 2020 it was time for full redevelopment. When it was dismantled it was found to be a perfect spider condominium, complete with pantries of stored food (bottom right), including more than the occasional bee.

An example of a 'population census' for 2011: Red: Osmia leaiana. *Yellow:* Heriades truncorum. *White:* Hylaeus *species; Green: solitary wasps.*

The Trellick Bee Tower was used by:

Osmia bicornis, O. leaiana, O. caerulescens, Heriades truncorum, Megachile willughbiella, M. ligniseca, Hylaeus communis, H. signatus, Anthidium manicatum *and the wasps* Sapyga quinquepunctata, Gasteruption jaculator, G. assectator, Dipogon subintermedius *and* Ephialtes manifestator.

The Bee Shard, March 2020

Developed to replace the Trellick Bee Tower (in place from 2010-2020).

Structure of treated timber painted with garden furniture paint. Not in direct contact with bee blocks.

Roof, lined with old pond liner; slopes backwards so that rain drains down the back of the structure.

Floors made of untreated, scaffolding planks, re-used from old Trellick Bee Tower (more than 10 years old). Each floor overhangs bee blocks below and slopes down from back to front, to throw rain forward away from the blocks.

New nesting blocks of untreated softwood drilled with holes with diameters of 9mm, 8mm, 6mm, 4mm and 2.5mm. The minimum hole depth is 100mm. Others are as deep as the timber allows. The holes were smoothed at the rim with a countersinking tool; running the drill in reverse reduced any rough interior walls. All waste from inside was knocked out.

Old blocks from the Trellick Bee Tower containing live nests awaiting emergence. These were moved into the shade at the rear of the tower before they could be reused by new females. Space then filled with old oak blocks.

Oak blocks at base yet to be drilled with deeper holes for larger leaf-cutter and red mason bees.

Sturdy timber block to give weight at base (not drilled); rests on concrete blocks partially sunk into the ground.

The Bee Shard is firmly secured with an additional 'spine' at the rear that is sunk into the ground. There are fewer holes than the Trellick Bee Tower, which means they are in greater demand. The Bee Shard was installed in early March 2020 and was quickly inhabited by red mason bees (as was its predecessor in 2010). Because of the Covid-19 pandemic the arrival of other species is unknown, but by June 2020 there were *Osmia leaiana, O. caerulescens, Heriades truncorum* and *Hylaeus nests, Gasteruption jaculator* patrolling and *Megachile willughbiella* males hanging about.

The classification of organisms – taxonomy – is a science considered under pressure from lack of funding and training of practitioners, at a time of increasing recognition of the loss of biodiversity. The lack of knowledge of that diversity prompted the setting up of the Global Taxonomy Initiative by the Convention on Biological Diversity. How we name and classify species is also in flux as techniques evolve and as we apply these new techniques to recognised groups of organisms (by splitting species, for example) and to the constant flow of discoveries of new species (pp.5, 46). Phylogenetic techniques using DNA sequencing have rapidly improved our understanding of species relationships and their evolution. They are providing alternative approaches to interpreting the 'family tree' of the bees and increasing our sophistication and accuracy in applying the Linnean system of classification (named after the 18th-century Swedish zoologist Carl Linnaeus). Using this binomial system a recognised species is given a name consisting of two parts (as in *Anthophora plumipes*), using Latin as the basis for this universal taxonomic language. The first part of the name is the genus (plural 'genera'), the second part is the species.

This sketch (right) is of a female bee of the genus *Anthophora* and species *plumipes*. These Latinised species names are written in italics with a capital for the genus name and lower case for the species name. If a species is referred to repeatedly in a text the genus name will often be abbreviated to its capital letter. Genera and species are then grouped into increasingly higher levels of taxonomic rank. There are many subdivisions, such as sub-family/family/super-family. The full taxonomy (as of 2020) for *Anthophora plumipes* is shown in the box opposite.

The 'family vine' shown on page 6 was drawn with reference to this arrangement, with *Anthophora*, for example, grouped into the tribe Anthophorini (note the suffix –ini for a 'tribe'), into the sub-family Apinae (suffix –inae) and then into the family Apidae (suffix –idae). All other bee genera are arranged into families such as Megachilidae, Colletidae and Halictidae. In some older books, written before molecular studies expanded, you might find that *Anthophora* bees are given their own family, the Anthophoridae. A current alternative classification, used in the *Handbook of the Bees of the British Isles*, makes Apidae the over-arching bee family with Melittidae, Megachilidae, Andrenidae, Halictidae and Colletidae demoted to sub-family status alongside the Apinae. Compare the two classifications of *Osmia bicornis* below:

Hymenoptera
Apoidea
Megachilidae
(leaf-cutter/ mason bees)
Megachilinae
Osmiini
Osmia
Osmia bicornis

Red mason bee

Hymenoptera
Apoidea
Apidae
(all bees)
Megachilinae
(leaf-cutter/mason bees)
Osmiini
Osmia
Osmia bicornis

Rank	Name	Description
Kingdom:	**Animalia**	**(animals, Metazoa)**
Subkingdom:	Bilateria	(animals with bilateral symmetry)
Infrakingdom:	Protostomia	(mouth is the first opening developing in the embryo)
Superphylum:	Ecdysozoa	(animals with exoskeletons that are moulted)
Phylum:	**Arthropoda**	**(animal with exoskeleton, segmented body, jointed appendages)**
Subphylum:	Hexapoda	(six-legged arthropods)
Class:	**Insecta**	**(insects)**
Subclass:	Pterygota	(insects with wings)
Infraclass:	Neoptera	(modern insects with folding wings)
Superorder:	Holometabola	(with full metamorphosis)
Order:	**Hymenoptera**	**(ants, bees and wasps)**
Suborder:	Apocrita	(with a 'wasp waist')
Infraorder:	Aculeata	(ovipositor modified into a sting)
Superfamily:	Apoidea	(bees, sphecoid wasps, apoid wasps)
Family:	**Apidae**	**(bumble bees, honey bees, stingless bees)**
Subfamily: Apinae		
Tribe:	Anthophorini	
Genus:	***Anthophora*** Latreille, 1803	
Species:	***Anthophora plumipes*** (Pallas, 1772)	

The names after the genus and species names refer to the first scientist/discoverer (with date) to describe the organism.

Until recently, every-day names in English for bees were only established for the more frequently experienced species such as the red mason bee or hairy-footed flower bee. Many more are now in accepted use and Falk's field guide uses the following English names for the bees in this book:

Andrena bicolor	Gwynne's mining bee
A. clarkella	Clarke's mining bee
A. cineraria	Ashy mining bee
A. florea	Bryony mining bee
A. fucata	Painted mining bee
A. fulva	Tawny mining bee
A. haemorrhoa	Orange-tailed mining bee
A. nigroaenea	Buffish mining bee
A. nitida	Grey-patched mining bee
A. nigrospina	Scarce black mining bee
A. pilipes	Black mining bee
A. ruficrus	Northern mining bee
A. tarsata	Tormentil mining bee
Anthidium manicatum	Wool carder bee
A. septemspinosum	Seven-spined wool carder bee
Anthophora quadrimaculata	Four-banded flower bee
A. bimaculata	Green-eyed flower bee
Apis mellifera	Honeybee
Bombus distinguendus	Great yellow bumblebee
B. hortorum	Garden bumblebee
B. hypnorum	Tree bumblebee
B. lucorum	White-tailed bumblebee
B. pascuorum	Common carder bee
B. terrestris	Buff-tailed bumblebee
Coelioxys elongata	Dull-vented sharp-tail bee
C. inermis	Shiny-vented coelioxys
C. quadridentata	Grooved sharp-tail bee
Colletes daviesanus	Davies' colletes
C. floralis	Northern colletes
C. hederae	Ivy bee
Dasypoda hirtipes	Hairy-legged mining bee
Halictus quadricinctus	Giant furrow bee
H. rubicundus	Orange-legged sweat bee
H. scabiosae	Great banded furrow bee
H. tumulorum	Bronze sweat bee
Heriades truncorum	Large-headed resin bee
Hylaeus communis	Common yellow-face bee
H. hyalinatus	Hairy yellow-face bee
H. signatus	Mignonette yellow-face bee

Lasioglossum calceatum	Common furrow bee
L.morio	Green furrow bee
L. minutissimum	Least furrow bee
Megachile centuncularis	Patchwork leaf-cutter bee
M. ligniseca	Wood-carving leaf-cutter bee
M. willughbiella	Willughby's leaf-cutter bee
Melecta albifrons	Round-spotted melecta
Nomada fabriciana	Fabricius' nomad
N. flava	Flavous nomad
N. goodeniana	Gooden's nomad
N. leucophthalma	Early nomad bee
N. marshamella	Marsham's nomad
N. panzeri	Panzer's nomad
Osmia bicornis (syn. *O. rufa*)	Red mason bee
O. caerulescens	Blue mason bee
O. cornuta	European orchard mason bee
O. inermis	Mountain mason bee
O. leaiana	Orange-vented mason bee
O. parietina	Wall mason bee
O. uncinata	Pinewood mason bee
Sphecodes gibbus	Dark-winged blood bee
Stelis breviuscula	Little dark bee
Xylocopa violacea	Violet carpenter bee

This lovely female wool carder bee was licking my finger tip with her glossa, presumably attracted to the salts in my sweat.

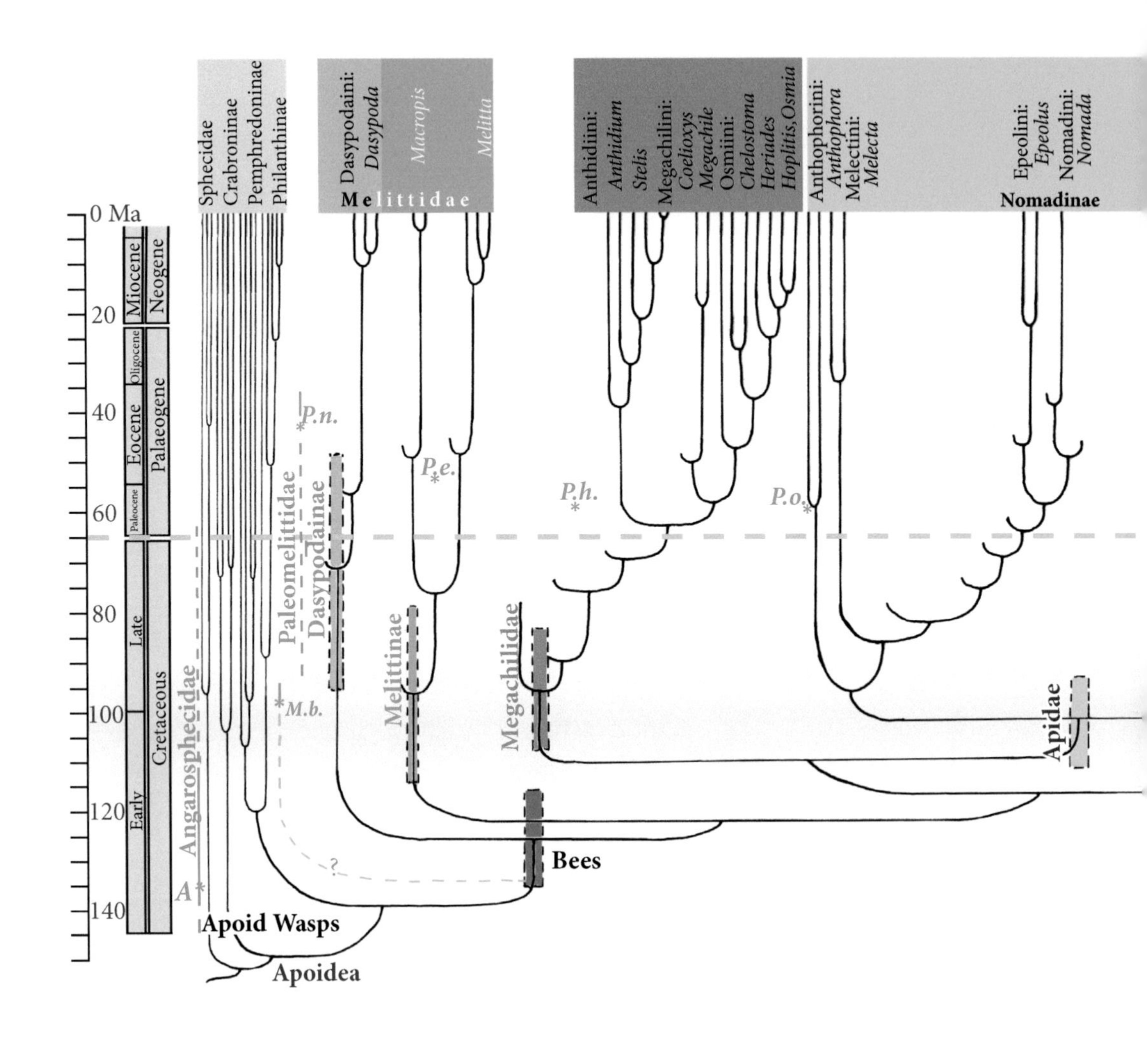

This phylogeny (an evolutionary family tree) for bees is based on that compiled by Cardinal and Danforth in 2013. Only lines represented by genera in Great Britain and Ireland are fully shown. Some older fossil bees (light blue) are added from a study of the Anthophorine ancestor *Paleohabropoda* by Michez et al. (2009) and from research by Ohl and Engel (2007).

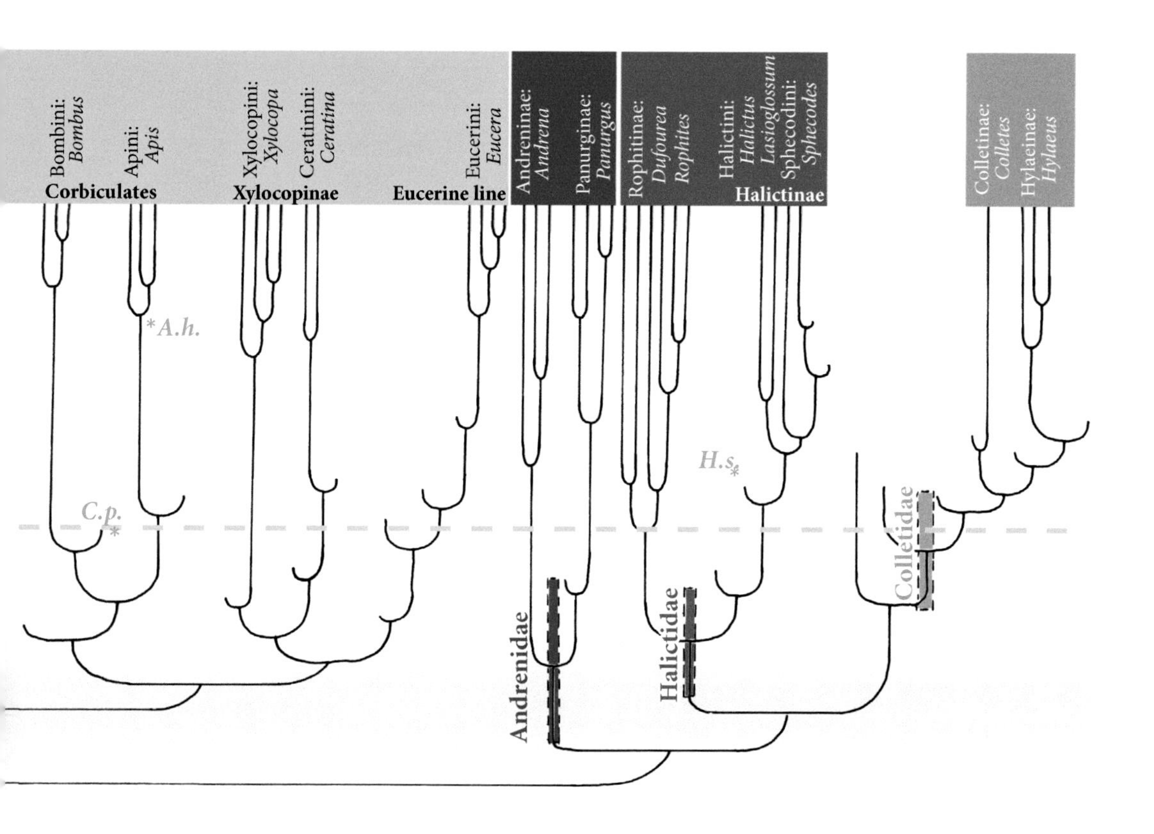

On the left are the apoid wasps (grey box), the closest related Hymenopterans in the superfamily Apoidea to bees. There was a split between wasps and bees in the early Cretaceous period. Coloured vertical bars indicate age ranges for the origins of bees (113–132 MA) and the major living bee families. All these emerged before the mass extinction at the end of the Cretaceous period (the boundary is shown as a grey dotted line) and during the evolutionary radiation of angiosperms (flowering plants), which is depicted by the misty yellow band across the diagram in the middle Cretaceous period. A selection of fossil Hymenoptera (some shown in p.47–50) are in light blue: *A: Angarosphex magnus; M.b.: Melittosphex burmensis; P.n.: Paleomelitta nigripennis; P.e: Paleomacropis eocenicus; P.o.: Paleohabropoda oudardi; C.p.: Cretotrigona prisca; A.h.: Apis henshawi; H.s.: Halictus? saveneyi.* (The identity of the last fossil is unconfirmed).

Abdomen: the hind-most of the three parts of the insect body. In bees it strictly includes the propodeum, attached to the thorax and separated from the rest of the abdomen (known as the *gaster*) by the *petiole.*

Aculeate: A member of the Aculeata, the group of insects comprising ants, bees and wasps, possessing a sting.

Aestivation: a period of dormancy (*diapause*) used to survive a hot or dry period.

Aggregation: in bees, a concentration of nests in a favoured site, often referring to mining bee nest holes.

Amber: hard, translucent, fossilised tree resin, often containing insect remains. There are important deposits of amber for fossilised bees in, for example, Myanmar, the Baltic and the Dominican Republic.

Angiosperm: a plant that reproduces using flowers bearing the reproductive organs and cells with the seeds enclosed within an ovary and maturing into fruit (herbaceous plants, grasses, shrubs, most trees not including conifers whose seeds are unprotected by a fruit – these are gymnosperms).

Antenna(e): segmented multi-sensory appendages arising from the head of an insect.

Anther: in the flower, the part of the stamen that produces pollen.

Anthropocene: term for the current geological period in which human activity has been the dominant influence on climate and the environment, still debated but becoming more widely used.

Aorta: the main artery of the body; in insects the tube that serves to transport nutrients via hemolymph to the head.

Apoid: insect belonging to the superfamily Apoidea – bees and allies.

Apoptosis: the death of a cell to be expected during an organism's growth: 'programmed cell death'.

Atrium: here, the space just inside the abdominal spiracles where air enters the bee's body and which has filter hairs to remove unwanted debris from the air.

Auricle: in corbiculate bees this is a projecting shelf or rod on the *basitarsus* with a fringe of hairs that serve to collect pollen brushed from the opposite leg and then to push and press it into the concave surface of the *tibia* (the *corbicula*). A part of the 'pollen press' in honeybees and bumblebees.

Basitarsus: see tarsus.

Carbohydrate: group of organic compounds including sugars, starches and cellulose containing carbon, hydrogen and oxygen, used in organisms as structural material and energy storage.

Caste (and 'sub-caste'): a group of individuals within a social species that has become behaviourally, and often anatomically, distinct from others in the colony for particular tasks. Generally there are three castes: queen, worker and drone (male), though some ants and termites have sub-castes of workers for different functions (eg 'soldiers').

Central body: part of the multifunctional central complex of the insect brain, sensitive to polarised light and responsible for navigational integration, the use of the sun as a compass and landmark memory.

Cephalothorax: a fused head and thorax, as in spiders. Here it occurs in the female parasite *Stylops*, embedded between the tergites of a bee with the cephalothorax projecting.

Chemoreceptor: a sensory cell, body part or organ responsive to chemical stimuli e.g. on antennae.

Chitin: a tough, fibrous, often translucent material that is the major component of the cuticle of the exoskeleton of an insect.

Chorion: the membrane surrounding the egg of an insect including a waxy layer to prevent water loss.

Chromosome: a thread-like structure (the spiral double helix) of nucleic acids (DNA) and protein, that carries genetic information (genes) in the nucleus of most living cells.

Cibarium: the first cavity of the bee's digestive tract into which food is pumped by muscles in its wall, then passing the liquid on into the muscular pharynx.

Cladogram: a diagram showing relative evolutionary relationships between species based on shared characteristics.

Clavate: club-shaped (the abdomen of the male large-headed resin bee, *Heriades truncorum*).

Cleptoparasitism: when a species feeds off the food or provisions of another species – a cuckoo. This may, or may not, lead to the death of the host species or its young.

Clypeus: a cuticular plate at the front of the head and at the base of the mouthparts, joined to the labrum.

Cob: a mixture of clay, mud, straw and often horse or cattle dung compressed and moulded to form walls.

Commensalism: an association between species that benefits one organism in the relationship but when the other takes neither harm nor benefit.

Compound eye: an eye with multiple facets or ommatidia, in insects and crustaceans.

Corbicula(e): a smooth, often shiny concavity fringed with hairs on the hind tibiae of honeybees and bumblebees (the corbiculate bees) used for collecting a compacted pollen/nectar mix to return to the nest. A similar structure on the propodeum of some *Andrena* species can be called a corbicula.

Corolla: the petals of a flower enclosing the reproductive organs and supported/surrounded by the sepals.

Coxa: the first segment of the leg, with very limited movement, between body and *trochanter.*

Cuckoo: in birds, a species laying its eggs in the nest of another species: the host adults rear the cuckoo young; in bees, the eggs are laid in the nest or cells of another species; a *cleptoparasite.*

Cuticle: a hard, waxy, protective layer consisting of linked plates on the outer surface of an invertebrate composed mainly of *chitin.*

Diapause: a period of suspension of metabolism and development in insects to cope with adverse environmental conditions. In solitary bees often during the pre-pupal stage.

Dichotomous keys: an identification key consisting of a series of numbered twin statements from which a choice is made to match the sample, the choice leading to another couplet and so on deeper into the key, until the final choice is made giving the id. They require pinned samples carefully collected and mounted and usually a lens or microscope to follow all the steps.

Diploid: a cell with two sets of *chromosomes*, one from each parent. See *haploid.*

Dorsal: the top surface of the body, facing you when viewing a resting insect from above.

Dufour's gland: an important gland in the abdomen, the secretions of which play a variety of roles such as waterproof nest cell lining, contact pheromones and larval food supplements.

Ecdysis: the process of shedding the outer cuticle during moulting from one life stage to another.

Ectotherm/Endotherm/Heterotherm: Ecto- dependent on external sources of heat to raise body temperature; **Endo**- capable of/dependent on internal generation of body heat, can maintain body temperature above the surrounding environment; **Hetero**- shifting from self-regulation of body temperature to allowing the environment to affect it, examples are bumblebees and hummingbirds (the latter entering torpor during cold nights to save the metabolic cost of maintaining body temperature).

Endoparasite: a parasite living within the body of its host; Endoparasitoid: an organism living within a host but eventually killing the host.

Endoskeleton: an internal skeleton as in the bony or cartilaginous skeletons of vertebrates.

Endosymbiosis: a type of *symbiosis* in which one of the symbiotic organisms lives within the body of another.

Enzyme: a substance, often a protein, produced by an organism that acts to enhance or stimulate a particular biochemical reaction within the organism.

Epidermis: the outermost, surface layer of the skin or exoskeleton.

Espalier: a method of training (particularly fruit-) trees to form horizontal layered branches (laterals) either side of a central stem.

Eusocial (and 'advanced'/ 'primitively eusocial'): advanced social organisation; a group of related individuals (often unfertilised/sterile females) from different generations cooperate to rear the young of one or more 'queens'. Different levels of eusociality are often identified: 'advanced' or 'primitive'.

Exoskeleton: an external supporting skeleton often a rigid 'shell' or a system of interlinked plates as in insects and crustaceans.

Facultative: an optional condition, operating under some circumstances but not others; contrasted with obligate.

Femur: the part of the leg between the *trochanter* and the *tibia.*

Fibula: in bees this is a small rod projecting from the *tibia* of the front leg, adjacent to the cleaning notch on the *basitarsus*, used for gripping the antenna within the notch for cleaning.

Flabellum: in bees, a spoon-like lobe at the tip of the *glossa*; sometime 'labellum'.

Flagellum (-a): the long, articulated part of the antenna; in bees, consisting of the final 10 antennal segments in the female, the final 11 segments in the male.

Floccus (flocci): a tuft of curled hairs on the hind *trochanters* of bees, especially in, for example *Andrena* species. Improves the efficiency of the pollen gathering *scopa.*

Foundress: a fertilised female initiating a nest, used especially in social insects to distinguish from later females that may become workers or later unfertilised females yet to become foundress queens.

Galea: an extended lobe of the *maxilla*, a pair of which (**galeae**) surround, partially protect, and form a tube about, the *glossa* of bees.

Ganglion (-a): in insects, a cluster of several nerve cells (neurons); the ganglia are arranged along the *ventral* nerve chord, usually one per segment.

Gaster: the part of the *abdomen* posterior to the waist (*petiole*) – ie not including the *propodeum.*

Genus (Genera): in taxonomy, the group that ranks above the *species* but below the family; given a capital letter in the latin name of the organism.

Germ cell: a cell containing half the number of *chromosomes* of a *somatic* cell; can unite with a germ cell from an individual of the opposite sex of the same species to form a new individual.

Glossa: in bees a long lobe of the mouthparts, part of the *labium*, the 'tongue' or proboscis.

Guild: A concept in ecology to cover a group of unrelated species exploiting the same set of resources in a similar way eg a 'pollinator guild': bee/hoverfly/butterfly species together using meadow flowers to a similar degree; or a 'nesting guild' – a set of species using the same habitat or ground material as nest sites.

Gyne: the reproductive female in eusocial species – usually known as a queen.

Hamuli: the hooks arranged along part of the front edge of the hind wing of Hymenoptera that link with the adjacent rear edge of the fore wing so joining the wings in flight to act as one aerofoil.

Haploid: having one set of chromosomes from only one parent.

Haplodiploidy: a type of sex-determination in insects: females are produced from fertilised, *diploid* eggs, males from unfertilised, haploid eggs.

Hemimetabolism: development that has stages – *instars* – that are smaller versions of the adult minus genitalia and wings – a form of "incomplete metamorphosis". See *nymph*.

Hemolymph: the 'blood' of insects.

Heterogeneous: diverse in character or content.

Hibernation: a period of dormancy (*diapause*) used to survive winter.

Histoblasts: an embryonic cell that forms tissue.

Holometabolism: development after hatching of the egg that has three distinct stages: the larval, pupal, and finally adult (*imago*) stage – often termed "complete metamorphosis".

Homogeneous: consisting of parts that are all of the same type.

Hypopharynx/ Hypopharyngeal glands: the **hypopharynx** is a small lobe in the centre of the mouth of an insect forming the floor of the *cibarium*. The paired **hypopharyngeal glands** open onto the front (upper) surface of the hypopharynx, the *head salivary glands* open below the hypopharynx.

Imaginal disc: a disc-like thickening of cells in the *epidermis* of the larva that contains specialist cells that will go on to form a particular part of the adult anatomy at pupation.

Imago: the final, adult stage of an insect, often winged.

Inquiline: a species sharing the living space of another species but without having any impact on its neighbour(s).

Instar: a larval (or nymphal) stage between moults.

Juvenile hormone: a hormone in insects that prevents moulting.

Labial palp: one of a pair of sensory appendages on the labium of insects – part of the proboscis; in bees elongated with 5 segments that, with the *galeae* surround the *glossa* to form the food canal.

Labium: the structures on the ventral surface of the mouth forming the hairy "tongue" of the bee (comprises the postmentum, the *prementum*, the *glossa* and the *labial palps*). With the *maxillae* the labium forms the proboscis of the bee.

Labrum: a plate on the upper (dorsal) surface of the mouthparts, sometimes distinctive in bees (and frequently pale coloured in male solitary bees).

Lancet: the serious 'business end' of the bee sting; paired and barbed in honeybees, it penetrates skin, then in honeybees works deep and lodges into the skin. See also *stylet.*

Larva(e): the immature stage of *holometabolous* insects, typically a relatively simple, soft-bodied form with reduced appendages, head and mouthparts and *imaginal discs* ready for pupation.

Machair: low-lying land near the coast formed by the deposition of sand and shell fragments by the wind; in Scotland (particularly the Hebrides) and Ireland. Used for traditional grazing or arable crops and often with a rich and diverse flora. Extremely valuable habitat for many species of invertebrate and nesting birds.

Mandible: mouthpart used for grasping, biting or cutting; the jaw of the insect.

Maxilla (-e): the paired structures of the mouth of the bee behind the mandibles and comprising the stipes (singular, plural stipites), an extension of which forms the *galeae*, and the maxillary palps, one each side of the head (the palps are reduced in size in bees).

Mechanoreceptor: sensory organ or cell that responds to mechanical stimuli such as touch or sound.

Metamorphosis: in an insect the process of transformation from an immature form to the adult through two or more distinct stages.

Microbial: pertaining to a micro-organism, particularly to a bacterium causing disease.

Monolectic: feeding from one species of flower.

Mulch: material spread over a soil surface to insulate the ground and/or introduce nutrients and organic matter: leaf litter, bark or wood chips, compost or manure.

Mushroom bodies: the parts of the bee brain playing a key role in learning and memory.

Nectary: a gland secreting nectar (sugar-rich fluid) to encourage pollinators to visit the plant; the gland can be within the flower or in some plants on leaves or stems.

Neophyte: a species new to an environment or region but now an established part of the flora/fauna.

Nymph: the immature stage of a *hemimetabolous* insect that is a smaller version of the adult without wings or genitalia.

Obligate: a condition operating under all circumstances, contrasted with *facultative.*

Ocellus (-i): small, lens-like, light sensitive structure; three arranged in a triangle between the two main compound eyes; "simple" eyes.

Oesophagus: the main part of the alimentary canal linking the *pharynx* to the crop (in the honeybee the latter is often called the honey sac).

Olfactory: relating to the sense of smell.

Oligolectic: feeding from a limited set of flowers of plants that are in the same genus or are closely related.

Ommatidium (-a): a unit of the compound eye comprising an outer lens, a crystalline cone and a column of photoreceptor cells with screening pigment cells to prevent the entry of stray light.

Operculum: a structure that closes or covers an opening, here found at the opening to the first spiracle on the thorax of the bee.

Optic flow: the apparent visual motion experienced by the eye as the observer moves through a scene of static objects. As the bee flies above an object the image of the object constantly changes with respect to the part of the compound eyes stimulated enabling the brain to interpret flight speed and relative positions of objects.

Ovary: female reproductive organ (in pairs), in which eggs are produced.

Ovipositor: an egg-laying appendage of the female insect, adapted in some Hymenoptera to create a long and flexible apparatus; the structure from which the sting evolved.

Parasite: an organism exploiting another organism, living on or in the host, harming it but generally not killing it – see *parasitoid.*

Parasitoid: an organism exploiting another organism, living on or in the host, generally killing it – see *parasite.*

Pedicel: the segment of the antenna between the basal *scape* and the flexible *flagellum.* The apparent 'hinge' of the antenna.

Petiole: a narrow attachment point: the narrow constriction between the *propodeum* on the *thorax* and the abdominal segments (the *gaster*): the 'wasp waist'.

Pharynx: the muscular part of the mouth of the bee behind the *cibarium* and before the *oesophagus.*

Phenology: the study of cyclic/seasonal phenomena; the relationships between fauna, flora and climate, especially now with respect to climate change.

Pheromone: a chemical released by an animal to the air, or laid on substrate, the presence of which affects the behaviour of others of the species, sometimes between species: a form of chemical communication.

Phylogenetics/phylogeny: the study of evolutionary relationships between organisms (species, individuals or genes) and the construction of an evolutionary model for the group. The phylogeny shows these relationships with respect to a geological time scale.

Plasticity ('social plasticity'): a flexible response to environmental conditions in terms of social organisation in a species.

Plumose: hairs with fine secondary filaments or branches giving a feathery appearance.

Polarised light: un-polarised light propagates as waves at right angles to the direction of travel, but in any plane with respect to that direction. If light passes through a filter, however, the vibrating waves can be restricted to a single plane: linear polarisation. Circular and elliptical polarisaton can also occur. Scattering of light in the environment leads to naturally occurring polarisation. Animals (such as bees) that are able to detect this polarisation can, for example, detect the position of the sun when it is not otherwise visible or in overcast conditions and hence continue to navigate in those conditions.

Polylectic: feeding from a wide range of unrelated flowers: the generalist feeder.

Polymorphic: occurring in several different forms within the one species.

Prementum: the part of the *labium* between the postmentum and the *glossa*; part of the 'tongue' of the bee.

Prepupa(e): the mature state of the larva after it has defecated but before full pupation. There is often a period of *diapause*, temporary cessation of development, as a prepupa.

Pronymph: in hemimetabolous insects a short, non-feeding stage with simpler morphology, between hatching and the first *instar nymph.*

Propodeum: the first abdominal segment that in ants, bees and wasps, is fused with the thorax and separated from the rest of the abdomen by the 'wasp waist'.

Protandrous: the male reproductive organs (in hermaphrodites and many plants) come to maturity before the female. Also used for bees whose males emerge earlier than females.

Punctate: bearing punctures – small, usually circular depressions in the cuticle.

Pupa(e): the final stage of transformation from larva to adult. The larval body is broken down and reorganised (see *imaginal disc*). The pupa is immobile and vulnerable in this time of change. Most solitary bees do not over-winter as a pupa.

Puparium (-a): a pupal case.

Pygidium: a raised, often triangular, central area on the last visible tergite (6) used by female solitary bees for smoothing the nest walls and removing debris.

Rima: central groove or furrow with microscopic hairs running down tergite 5 of female *Halictus* and *Lasioglossum* bees.

Salivary glands: there are two pairs in bees – one pair in the head and one in the *thorax*. Secretions from both pairs emerge in the mouth below the *hypopharynx*. Secretions are involved with food softening, lubrication of chewed material and in grooming.

Scape: the first segment of the *antenna* before the *pedicle*. Often elongated and sometimes swollen.

Scopa(e): a dense patch of hairs usually on the hind legs or the underside of the abdomen of female bees, adapted to collect and carry pollen back to the nest.

Sepal: one of several of the outermost parts of the flower resembling leaves, enclosing the reproductive organs and the petals before opening.

Soma/somatic cell: the parts of an organism other than the reproductive cells (see *germ cell*)

Sonication: in bees, the use of sound vibrations of particular frequencies to assist in loosening and releasing pollen from some flower species: "buzz pollination".

Species: a fundamental taxonomic unit, ranking beneath the *genus*. A grouping of individuals with similar (and unique) characteristics, capable of exchanging genes and reproducing to produce viable offspring. Can be split into subspecies.

Spermatheca: an internal storage unit in the female for holding and keeping sperm alive.

Spiracle: small, often valved, openings, in pairs, along the sides of the thorax and abdomen in insects allowing air to enter the *tracheae*.

Stamen: the male fertilising organ of a flower consisting of a filament bearing the anthers.

Sternite: a ventral plate of the insect body (in bees, generally referring to the *abdomen* or *gaster*).

Steroids: organic compounds with 17 carbon atoms arranged in four rings, including many hormones, alkaloids and vitamins.

Stigma: the point at the tip of the style to which pollen sticks, to then germinate and enter the style.

Style: the narrow, elongated structure between ovary and stigma.

Stylet: part of the bee sting attached to the venom bulb – a needle-like structure. See also *lancet*.

Stylopisation, Stylopised: a bee that is hosting the parasite *Stylops* is said to be stylopised.

Subsocial: one of the terms used to describe variations in social behaviour in the bees. Females remain to feed or defend the young that are not their own, though without being infertile themselves and without full-blown eusociality.

Superorganism: a term adopted by W. Morton Wheeler and developed by Edward O. Wilson and others to characterise the Hymenopteran eusocial nest as, effectively, an organism in its own right.

Symbiosis: more than one species living in close association, to the advantage of all. Symbiont: a species living in close association with another; adj. symbiotic.

Tarsus/Basitarsus/ Tarsomeres/pre-tarsus; plural tarsi: the tarsus is the lowest part of the leg below the tibia made up of 5 segments. The basitarsus is the first, upper most and longest segment, followed by three small tarsomeres and finally the pre-tarsus or foot of the bee.

Tergite: a dorsal segment of the abdomen of the bee. See *sternite*.

Thorax: the middle of the three main sections of the insect body between head and abdomen bearing wings, legs and their musculature. In bees the rear of the thorax – the *propodeum* – is technically part of the abdomen.

Tibia: the 4th section of the leg between femur and tarsus, often long and slender and bearing spines and spurs. In honeybees and bumblebees it bears the *corbicula*.

Trachea(e): fine tubes within the internal spaces of the insect body that transport oxygen and carbon dioxide directly to and from tissues. They open from the *spiracles*.

Tracheole: a very fine *trachea*, the last dividing branch of the tracheae.

Triungulin (pl. triungula): the active, first *instar larva* of parasitic stylopids and meloid beetles.

Trochanter: the 2nd section (the first freely articulating segment) of the leg between the *coxa* and the *femur*.

Tubule: a minute tube.

Ventral: the underside of the body, facing the ground in a resting bee

The impulse to design this book came from the stories I've been telling children and adults over the last 25 years about how bees live and where they fit into our world. I took most of the photographs used whilst working on the *Wildlife Garden at Roots and Shoots* in Lambeth, central London. Others are from:

P.6: The hairy-legged mining bee *Dasypoda hirtipes* on the 'family vine', was taken at Roots and Shoots by Dwayne Senior for a piece in the Sunday Times, 5 September 2004.

P.46: Photos of the cob wall by Christopher Wren and of *Anthophora pueblo* nests by Michael Orr (see references below).

P.70: The image of the ivy bee *Colletes hederae* is by Matt Smith and on the BWARS website.

P.86: Photo of Flatford Wildlife Garden by Amy Ward, RSPB.

P.90: Ivy bee photos taken by the author near Margate on the north Kent coast.

Pp.97 & 167: Male *Anthophora quadrimaculata* on lavender, taken on platform C of Waterloo East railway station, London.

P.130: The forest of alexanders *Smyrnium olusatrum* (very good bee forage) is on the north coast of the Isle Of Grain, Kent. Steven Falk's thumb and forefinger hold a female *Andrena fucata.*

Pp.171–172: Photographs of the first Irish hairy-footed flower bees are by Martin Molloy, courtesy of Úna Fitzpatrick, National Biodiversity Data Centre, Ireland.

There are also several photographs taken by the author in gardens near Ballylifford, Northern Ireland, Maryport, Cumbria and Greenwich, London.

All the drawings are by the author and have been made from specimens, from life or from original photographs and video. Additional information for some drawings was taken from online images as indicated below. Some scientific diagrams and the drawings of fossils have also been derived from published material acknowledged below. Scientific papers referenced here can almost all be viewed on or downloaded from the internet. A few may be in abstract form only. Some are very technical, but the discussion/conclusions are often illuminating and worth spending time on.

Hairy legs from around the world: other Anthophora *species, male mid legs, left to right: A.* retusa *(now restricted to the south and east of the British Isles. It ranges from Sweden to Iberia, eastwards to China though it is declining globally); A.* dufourii *(occurs in the Mediterranean and east to Turkmenistan); A.* hispanica *(ranges from Iberia, across North Africa to the eastern Mediterranean). Other species' hairy feet are shown in the following pages of notes and references.*

Mid legs of *Anthophora* males from around the world have been redrawn from illustrations in **Brooks, R.W.** (1988), 'Systematics and Phylogeny of the Anthophorine Bees (Hymenoptera: Anthophoridae; Anthophorinae)', *University of Kansas science bulletin,* 53 (9): 436–575; and **Brooks, R.W.** (1999), 'Bees of the genus Anthophora Latreille 1803 (Hymenoptera Apidae Anthophorini) of the West Indies', *Tropical Zoology,* 12 (1): 105–124.

Pp.3–4: The non-aculeate members of the Apocrita are now put into the 'infraorder' Terebrantes, making the term parasitica obsolete. The group contains the ichneumon wasps, cynipoid wasps (gall wasps producing plant galls for larval development) and evanioid wasps, which includes *Gasteruption* (pp.94–95). The pie chart is redrawn from one in **Grimaldi, D. and Engel, M.S.** (2005), *Evolution of the Insects,* Cambridge University Press. The proportions should not be regarded as fixed. For a discussion of the diversity of parasitoid Hymenoptera compared with the diversity of the beetles see **Forbes, A. et al.** (2018), 'Quantifying the unquantifiable: why Hymenoptera, not Coleoptera, is the most speciose animal order', *BMC Ecology* 18:21. Or perhaps it is a fungal planet because of the crucial role of fungi in all ecosystems: **Merlin Sheldrake** (2020), *Entangled Life*, Bodley Head; 'Fungi: the new frontier', **BBC Radio 4**, www.bbc.co.uk/programmes/m00132xm.

Pp.5–6: The table of genera and numbers of species is compiled using **Falk, S.** (2015) *Field Guide to the Bees of Great Britain and Ireland*, Bloomsbury, and **Else, G.R. and Edwards, M.** (2018) *Handbook of the Bees of the British Isles*, The Ray Society.

For species status in Europe see **IUCN** (2014), *European Red List of Bees*, www.iucn.org/content/european-red-list-bees. The largest bee in the world, discovered by Alfred Russel Wallace, is a species of *Megachile* (formerly *Chalicodoma*) described in **Messer, A.C.** (1984), 'Chalicodoma pluto: The world's largest bee rediscovered living communally in termite nests (Hymenoptera: Megachilidae)', *Journal of the Kansas Entomological Society,* 57 (1): 165–168. The original specimen collected by Wallace is in the Hope Entomological Collections at the Oxford University Museum of Natural History (oumnh.ox.ac.uk/wonderful-diversity-bees). It was 're-rediscovered' in January 2019 (www.theguardian.com/environment/2019/feb/21/worlds-largest-bee-missing-for-38-years-found-in-indonesia). See also **Linnean Society** lecture at www.youtube.com/watch?v=772mHQ4TeY8.

Pp.5–6 contd.: New *Lasioglossum* species were described for Eastern USA in 2011: **Jason Gibbs** (2011), 'Revision of the metallic *Lasioglossum (Dialictus)* of eastern North America (Hymenoptera: Halictidae: Halictini)', *Zootaxa* 3073: 1–216; see also: blogs.ethz.ch/osmiini/category/additional-species.

For new species in Britain see **Notton, D., Tang, C. & Day, Anthony R.** (2016), 'Viper's Bugloss Mason Bee, *Hoplitis (Hoplitis) adunca*, new to Britain (Hymenoptera, Megachilidae, Megachilinae, Osmiini)', *British Journal of Entomology & Natural History* 29: 134-143. Also for *Nomada alboguttata* see www.bwars.com/bee/apidae/nomada-alboguttata.

For *N. facilis*: **Notton, D. & Norman, H.** (2017) 'Hawk's-beard Nomad Bee, *Nomada facilis*, new to Britain (Hymenoptera: Apidae)', *British Journal of Entomology & Natural History* 30: 201–214; *Osmia cornuta* was first seen by Dusty Gedge in 2014 near Blackheath, southeast London: www.lnhs.org.uk/index.php;/articles-british/477-new-bee-consolidating-its-foothold-in-london (there were further records in 2017 and annually since then).

For *Heriades rubicola*: **Cross, I. & Notton, D.** (2017) 'Small-Headed Resin Bee, *Heriades rubicola*, new to Britain (Hymenoptera: Megachilidae)', *British Journal of Entomology & Natural History* 30: 1–6.

Steven Falk (2015), in his *Field Guide to the Bees of Great Britain and Ireland,* refers to the possibility of *Andrena proxima* being two distinct forms in the UK. The split, introducing *Andrena ampla* to the British list, was confirmed in 2018.

Edwards, M., Hazlehurst, G., Wright, I. (2019), '*Stelis odontopyga* Noskiewicz (Hymenoptera: Megachilinae) new to Britain', *British Journal of Entomology & Natural History* 32:43–47.

New *Anthophora* species described by **Orr, M., Griswold, T., Pitts, J.P., Parker, F.D.** (2016), 'A new bee species that excavates sandstone nests', *Current Biology*, 26 (17): R779–R793 (described here on P.).

Orr. M, Pitts, J. and Griswold, T. (2018), 'Revision of the bee group *Anthophora* (Micranthophora) (Hymenoptera: Apidae), with notes on potential conservation concerns and a molecular phylogeny of the genus', *Zootaxa* 4511 (1): 1–193.

Mawlood, N.A., Amin, H.M. (2017), 'A new species of the bee *Anthophora* Latreille, 1803 (Hymenoptera: Apidae) from Kurdistan region – Iraq', *Kurdistan Journal of Applied Research.* 2 (1).

The 'family vine' was drawn with reference to the phylogeny in **Cardinal S. and Danforth B.N.** (2013), 'Bees diversified in the age of eudicots', *Proceedings of the Royal Society* B, 280: 20122686. For references on *Megalopta* vision see notes on the bee brain below.

A. dispar *(southern Europe, Iberia, north Africa, Israel, Syria).*

Pp.7–22: Diagrams in these sections were derived from the following references. These give more details and images of bee anatomy, especially:

Lesley Goodman (2003), *Form and Function in the Honey Bee*, pp.154–157, International Bee Research Association;

Ian Stell (2012), *Understanding Bee Anatomy: a full colour guide*, The Catford Press.

Mark L. Winston (1987), *The Biology of the Honey Bee*, Harvard University Press.

Stephen, W.P., Bohart, G.E., Torchio, P.F., (1969), 'The Biology and External Morphology of Bees', Oregon State University.

And, amazingly, given that it was first produced at the beginning of the 20th century and again in 1956, **Snodgrass, R.E.** (1910), 'The Anatomy of the Honey Bee,' U.S. Dept. of Agriculture, Washington. It is still available online and is worth studying.

Pp.9–10: Graham Stone (1993), 'Endothermy in the solitary bee *Anthophora plumipes*: independent measures of thermoregulatory ability, costs of warm-up and the role of body size', *Journal of Experimental Biology,* 174, 299–320 (1993).

Thermoregulation in honeybees and bumblebees is well covered in Lesley Goodman (2003), op. cit., pp.154–157, International Bee Research Association, in **Roberts, S. and Harrison, J.** (1999) 'Mechanisms of thermal stability during flight in the honey-bee *Apis mellifera*', *Journal of Experimental Biology* 202, 1523–1533, and in **Goulson, D.** (2003), *Bumblebees, Behaviour and Ecology*, Oxford University Press.

P.12: Specifically on glands: **Mitra, A.** (2013), 'Function of the Dufour's gland in solitary and social Hymenoptera', *Journal of Hymenoptera Research* 35, 33–58.

Katzav-Gozansky, T. Soroker, V., Hefetz, A., (2002), 'Honeybees Dufour's gland – idiosyncrasy of a new queen signal', *Apidologie* 33, 525–537.

Hefetz, A. (1987), 'The role of Dufour's gland secretions in bees', *Physiological Entomology,* 12, 243–253.

Dornhaus, A., Brockmann, A. & Chittka, L., (2003), 'Bumble bees alert to food with pheromone from tergal gland', *Journal Comparative Physiology A*, 189, 47–51.

Ayasse, M., Paxton, R.J., Tengö, J. (2001), 'Mating Behaviour and Chemical Communication in the Order Hymenoptera', *Annual Review of Entomology,* 46: 31–78.

P.14: The drawing of the bee brain was made with further reference to: **Menzel Group** (2012), 'The Virtual Atlas of the Honeybee Brain', www.bcp.fu-berlin.de/en/biologie/arbeitsgruppen/neurobiologie/ag_menzel/beebrain/index.html (Freie Universitat Berlin); **Smith, D.B., et al.** (2016), 'Exploring miniature insect brains using micro-CT scanning techniques', *Scientific Reports*, 6; 21768; Report on research by **Stanley Heinze** on *Megalopta*: 'New study shows how bee's brain functions to guide it home', www.sci-news.com/biology/bees-brain-guide-home-05303.html, and **Berry, R.P., Wcislo, W.T., Warrant, E.J.** (2011), 'Ocellar adaptations for dim light vision in a nocturnal bee', *Journal of Experimental Biology*, 214, 1283–1293.

P.14 contd.: On evolution of the brain: **Farris, S.M. and Schulmeister, S.** (2011), 'Parasitoidism, not sociality, is associated with the evolution of elaborate mushroom bodies in the brains of hymenopteran insects', *Proceedings of the Royal Society B*, 278, 940–951.

Smith, A.R., Seid, M.A., Jimenez, L.C., Wcislo, W.T. (2010), 'Socially induced brain development in a facultatively eusocial sweat bee *Megalopta genalis* (Halictidae)', *Proceedings of the Royal Society B*, 277, 2157–2163.

Lihoreau, M., Latty, T., Chittka, L. (2012), 'An exploration of the social brain hypothesis in insects', *Frontiers in Physiology* 3, Article 442.

O'Donnell, S., Bulova, S.J., DeLeon, S., Khodak, P., Miller, S., Sulger, E. (2015), 'Distributed cognition and social brains: reductions in mushroom body investment accompanied the origins of sociality in wasps (Hymenoptera: Vespidae)', *Proceedings of the Royal Society B*, 282: 20150791.

Whether bees can think in 'abstract' terms is worth debating: the honeybee dance language, communicating the nature of a floral resource using the sun's position on the horizon as a navigational aid in a dark hive (and adjusting for passage of time) has been described as a 'symbolic language'. **Lars Chittka**'s new book *The Mind of a Bee* (Princeton, 2022) is a beautifully presented review of research and ideas on bee cognition.

Pp.19–22: Further reading on animal flight and its evolution:

David E. Alexander (2002), *Nature's Flyers: Birds, Insects and the Biomechanics of Flight*, Johns Hopkins University Press & **David E. Alexander** (2015), *On the Wing: Insects, Pterosaurs, Birds, Bats and the Evolution of Animal Flight*, Oxford University Press, 2015.

Henk Tennekes (2009), *The Simple Science of Flight: from insects to jumbo jets*, Massachusetts Institute of Technology Press.

Robert Dudley (2000), *The Biomechanics of Insect Flight*, Princeton University Press.

David Grimaldi and Michael S. Engel (2005), *Evolution of the Insects*, Cambridge University Press. Bee energy expenditure from **Tom Seeley** (2010), *Honeybee Democracy*, Princeton University Press.

Pp.23–34: For good general texts on pollination and adaptation:

Proctor, M., Yeo, P. & Lack, A. (1996), *The Natural History of Pollination*, New Naturalist, Harper Collins. More recently, **Jeff Ollerton** (2021), *Pollinators & Pollination: Nature and Society*, Pelagic Publishing.

On the galeae: **Müller, A.** (1995), 'Morphological specialisations in central European bees for the uptake of pollen from flowers with anthers hidden in narrow corolla tubes (Hymenoptera: Apoidea)', *Entomologia Generalis*, 20(1–2):43–57. **Müller, A.** (1996), 'Convergent evolution of morphological specialisations in Central European bee and honey wasp species as an adaptation to the uptake of pollen from nototribic flowers (Hymenoptera, Apoidea and Masaridae)', *Biological Journal of the Linnean Society*, 57(3):235–252.

On *Ancyloscelis*: **Alves-dos-Santos, I. & Witmann, D.** (1999), 'The proboscis of the long-tongued *Ancyloscelis* bees (Anthophoridae/Apoidea), with remarks on flower visits and pollen collecting with the mouthparts', *Journal of the Kansas Entomological Society*, 72(3):277–288. These papers are discussed in **Danforth, Minckley and Neff** (2019), *The Solitary Bees, Biology, Evolution, Conservation*, Princeton University Press.

Explaining and drawing the honeybee pollen-collecting apparatus and technique **(p.27)** is a challenge – it is a four-dimensional phenomenon difficult to reduce to two! There are many other explanations in the honeybee literature – see Lesley Goodman (2003), op. cit. IBRA.

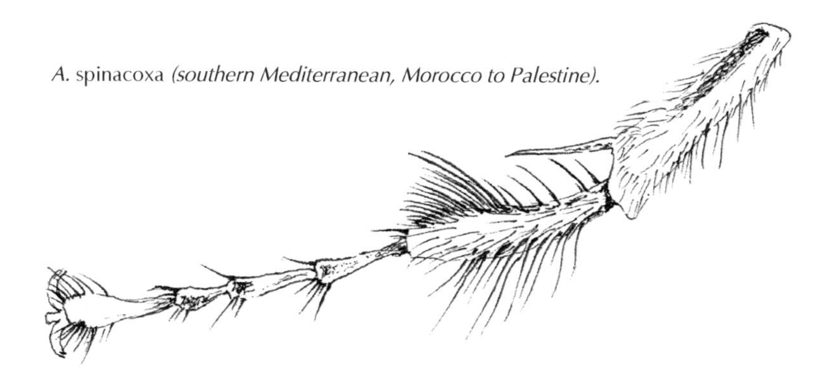
A. spinacoxa (southern Mediterranean, Morocco to Palestine).

The SEM image of the bumblebee corbicula was taken during a National Science Week public open day at Roots and Shoots – we borrowed a 'desktop' SEM for the weekend from JEOL UK Ltd (also **p.38**).

On bee corbiculae: **Charles D. Michener** (1999), 'The corbiculae of bees', *Apidologie*, 30 67–74.

With increased recognition of the importance of wild pollinators (as opposed to honeybees) and of the economic value of 'pollination services' to agriculture and horticulture, it is no surprise that there has been a lot of research in the last decade. However, it is also sobering to read that the problem has been recognised since at least 1935. Here is a quotation from *National Geographic Magazine* from April 1935 (by **J. I. Hambleton**): "In the pioneer stages of American agriculture, bumblebees and other native pollinating insects...were plentiful everywhere. But the planting of vast areas which once were forests, prairies and swamps with fields of grain, orchards and gardens upset the delicate balance of nature...Injurious species, afforded an enormous food supply, prospered and multiplied until now serious insect pests menace almost every important crop. Insecticides must be used to protect farm crops, particularly fruits. Unfortunately, these materials kill not only harmful but beneficial insects. The toll includes honeybees and other wild bees, as well as the efficient bumblebees – all the insects that carry pollen from one blossom to another. Even yet we scarcely realize the dependence of many plants upon insects to effect pollination." Now, 85 years later, we are still struggling to recognise the importance of wild pollinators and of not destroying them with chemicals.

Bryan Danforth's lab at Cornell University (https://www.danforthlab.entomology.cornell.edu/research/) has a long-term project on the pollination of crops, especially New York apples. All papers by Danforth (and his collaborators) can be viewed on the lab's website.

Here are some noteworthy papers from there and elsewhere:

Richards, A.J. (2001), 'Does low biodiversity resulting from modern agricultural practice affect crop pollination and yield?' *Annals of Botany*, 88, 165–172.

Park, M.G., Orr, M.C., Danforth, B.N. (2010), 'The role of native bees in apple pollination', *New York Fruit Quarterly*, 18 (1): 21–25.

Garibaldi, L.A., Steffan-Dewenter, I., Kremen, C., et al. (2011), 'Stability of pollination services decreases with isolation from natural areas despite honey bee visits', *Ecology Letters*, 14: 1062–1072.

Brittain, C., Williams, N., Kremen, C., Klein, A-M. (2013), 'Synergistic effects of non-Apis bees and honey bees for pollination services', *Proceedings of the Royal Society B*, 280: 20122767.

Park, M.G., Raguso, R.A., Losey, J.E., Danforth, B.N. (2016), 'Per-visit pollinator performance and regional importance of wild *Bombus* and *Andrena* (*Melandrena*) compared to the managed honey bee in New York apple orchards', *Apidologie* 47: 145–160.

Blitzer, E.J., Gibbs, J., Park, M.G., Danforth B.N., (2016), 'Pollination services for apple are dependent on diverse wild bee communities', *Agriculture, Ecosystems and Environment* 221: 1–7.

Lindström, S.A.M., Herbertsson, L., Rundlöf, M., Bommarco, R., Smith, H.G., (2016), 'Experimental evidence that honeybees depress wild insect densities in a flowering crop', *Proceedings of the Royal Society B* 283: 20161641.

Grab, H., Blitzer, E.J., Danforth, B.N., Loeb, G., Poveda, K. (2017), 'Temporally dependent pollinator competition and facilitation with mass flowering crops affects yield in co-blooming crops', *Scientific Reports* 7: 45296.

Russo, L., Park, M.G., Blitzer, E.J., Danforth, B.N. (2017), 'Flower handling behaviour and abundance determine the relative contribution of pollinators to seed set in apple orchards', *Agriculture, Ecosystems and Environment* 246: 102–108.

Russo, L., Danforth, B.N. (2017), 'Pollen preferences among the bee species visiting apple (Malus pumila) in New York', *Apidologie* 48(6): 806–820.

Garratt, M.P.D., Breeze, T.D., Jenner, N., Polce, C., Biesmeijer, J.C., Potts, S.G. (2014), 'Avoiding a bad apple: Insect pollination enhances fruit quality and economic value', *Agriculture, Ecosystems & Environment*, 184 (100): 34–40.

Garratt, M.P.D., Breeze, T.D., Boreux, V. et al. (2016), 'Apple pollination: demand depends on variety and supply depends on pollinator identity', *PLOS ONE*, https://doi.org/10.1371/journal.pone.0153889.

Pp. 29–30: For more on buzz pollination:

Callin Switzer (2014), 'Getting buzzed at the Arnold Arboretum', *Arnoldia* 71/4.

De Luca, P. & Vallejo-Marin, M. (2013), 'What's the buzz about? The ecology and evolutionary significance of buzz-pollination', *Current Opinion in Plant Biology*, 16: 1–7.

De Luca, P., Bussière, L.F., Souto-Vilaros, D., Goulson, D., Mason, A.C., Vallejo-Marin, M., (2013), 'Variability in bumblebee pollination buzzes affects the quantity of pollen released from flowers', *Oecologia*, 172: 805–816.

De Luca, P., Cox, D.A., Vallejo-Marin, M. (2014), 'Comparison of pollination and defensive buzzes in bumblebees indicates species-specific and context-dependent vibrations', *Naturwissen-schaften*, 101: 331–338.

Stephen L. Buchmann (1985), 'Bees use vibration to aid pollen collection from non-poricidal flowers', *Journal of the Kansas Entomological Society*, 58 (3): 517–525.

Corbet, S.A., Huang, S-Q., (2014), 'Buzz pollination in eight bumblebee-pollinated *Pedicularis* species: does it involve vibration-induced triboelectric charging of pollen grains?' *Annals of Botany,* 114: 1665–1674.

Morgan, T., Whitehorn, P., Lye, G.C., Vallejo-Marin, M. (2016), 'Floral sonication is an innate behaviour in bumblebees that can be fine-tuned with experience in manipulating flowers', *Journal of Insect Behaviour,* 29: 233–241.

Russell, A.L., Buchmann, S.L., Papaj, D.R. (2017), 'How a generalist bee achieves high efficiency of pollen collection on diverse floral resources', *Behavioural Ecology* 28 (4): 991–1003.

Whitehorn, P., Wallace, C., Vallejo-Marin, M. (2017), 'Neonicotinoid pesticide limits improvement in buzz pollination by bumblebees', *Scientific Reports* 7: 15562.

Cardinal, S., Buchmann, S.L., Russell, A., (2018), 'The evolution of floral sonication, a pollen foraging behaviour used by bees (Anthophila)', *Evolution*, 72–3: 590–600.

'Time and Motion':

Goulson, D., Hawson, J.A., Stout, J.C. (1998), 'Foraging bumblebees avoid flowers already visited by conspecifics or by other bumblebee species', *Animal Behaviour,* 55 (1): 199–206.

Gawleta, N., Zimmermann, Y., Eltz, T. (2005), 'Repellent foraging scent recognition across bee families', *Apidologie,* 36: 325–330.

Electro-reception:

Clarke, D., Whitney, H., Sutton, G., Robert, D. (2013), 'Detection and learning of floral electric fields by bumblebees', *ScienceExpress,* 21 Feb 2013.

Lihoreau, M., Raine, N.E. (2013), 'Bee positive: the importance of electroreception in pollinator cognitive ecology', *Frontiers in Psychology*, 4: 445.

Orchid pollination:

Ayasse, M., Dötterl, S. (2014), 'The role of preadaptations or evolutionary novelties for the evolution of sexually deceptive orchids', *New Phytologist*, 203: 710–712.

Schiestl, F.P., Ayasse, M. (2000), 'Post-mating odour in females of the solitary bee, *Andrena nigroaenea* (Apoidea, Andrenidae), inhibits male mating behavior', *Behavioral Ecology and Sociobiology*, 48: 303¬–307; Schiestl.

Schiestl, F.P., Ayasse, M. (2001), 'Post-pollination emission of a repellent compound in a sexually deceptive orchid: a new mechanism for maximising reproductive success?' *Oecologia* 126: 531–534.

Robbirt, K.M., Roberts, D.L., Hutchings, M.J., Davy, A.J. (2014), 'Potential disruption of pollination in a sexually deceptive orchid by climatic change', *Current Biology,* 24: 2845–2849.

Schiestl, F.P., Ayasse, M., Paulus, H.F., Löfstedt, C., Hansson, B.S., Ibarra, F., Francke, W. (2000), 'Sex pheromone mimicry in the early spider orchid (*Ophrys sphegodes*): patterns of hydrocarbons as the key mechanism for pollination by sexual deception', *Journal of Comparative Physiology,* 186: 567–574.

See also chapter 8 in **Christopher O'Toole** (2013), *Bees: a Natural History*, Firefly Books.

In addition to scent, colour, shape, electro-magnetism, we can also add sound to the complexity of plant-pollinator communication:

Veits, M., Khait, I., Obolski, U., Zinger, E., et al. (2018), 'Plants hear: Evening primrose flowers rapidly respond to the sound of a flying bee by producing sweeter nectar', bioRxiv: doi.org/10.1101/507319.

Pp.31-32: Suzuki, K. (1984), Pollination system and its significance on isolation and hybridization in JapaneseEpimedium (Berberidaceae). *Bot Mag Tokyo* 97, 381–396. https://doi.org/10.1007/BF02488670

Tang J, Quan Q-M, Chen J-Z, Wu T, Huang S-Q. (2019) Pollinator effectiveness and importance between female and male mining bee (*Andrena*). *Biol. Lett.* 15: 20190479. http://dx.doi.org/10.1098/rsbl.2019.0479

Pp.33–34: Observations made at Roots and Shoots and Blackheath, London.

Pp.35–36: The use of pheromones during mating is described in **Wittmann, D., Schindler, M., Blochtein, B., Barouz, D.** (2004), 'Mating in bees: how males hug their mates', *Proceedings of the 8th IBRA International Conference on Tropical Bees, Brazil*, 2004.

See also entry on BWARS website by **Nigel Jones**.

Also in **Graham Stone** (1995), 'Female foraging responses to sexual harassment in the solitary bee, *Anthophora plumipes*', *Animal Behaviour*, 50, 405–412.

Also: **Stone, G.N., Loder, P., Blackburn, T.** (1995), 'Foraging and courtship behaviour in males of the solitary bee *Anthophora plumipes*: thermal physiology and the roles of body size', *Ecological Entomology*, 20 (2), 169–183.

Robert J. Paxton (2005), 'Male mating behaviour and mating systems of bees: an overview', *Apidologie*, 36 (2): 145–156.

An excellent video of mating in *Anthophora plumipes* by **John Walters** in Devon in 2017 is on YouTube: https://www.youtube.com/user/jwentomologist/videos; https://www.youtube.com/watch?v=OyW6J3mCPys (last accessed Jan 2024).

Pp.37–38: The sensory role of hairs in bees is fully covered in **Lesley Goodman** (2003), op. cit. SEM photos here were taken at Roots and Shoots on a community open day; the drawing of an antenna tip is modified from a SEM image in Goodman.

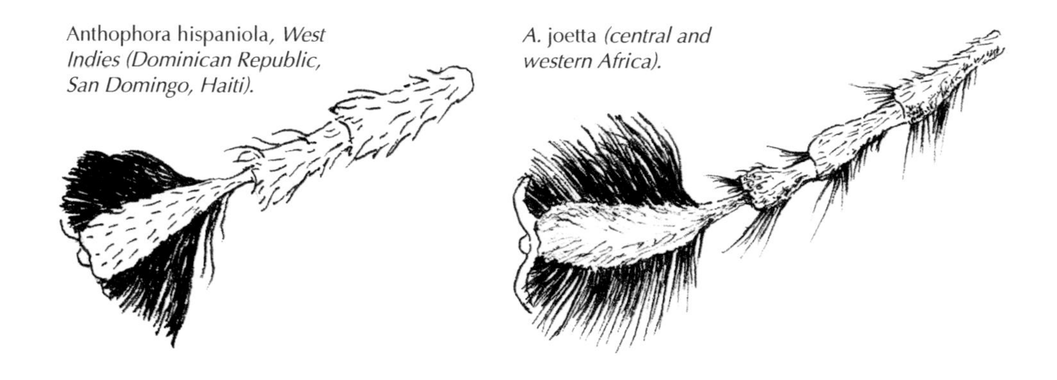

Pp.39–42: See **Lesley Goodman** (2003), op. cit. chapter 5; **Proctor, M., Yeo, P. & Lack, A.** (1996), op.cit.

Caffeine in nectar: **Wright, G.A., et. al.** (2013), 'Caffeine in floral nectar enhances a pollinator's memory of reward', *Science,* 339(6124): 120 (https://www.ncbi.nlm.nih.gov/pmc/articles/PMC4521368/).

Drawing of blue mason bee male partly based upon photos on BWARS website by **Josef Dvorak.**

Pp.43–46: Neff, J. L. (2008), 'Components of nest provisioning behavior in solitary bees (Hymenoptera: Apoidea)', *Apidologie,* 39: 30–45.

Megachile bees have been recorded using plastic agricultural waste as nest material: **Allasinoi, M.L. , Marrero, H.J. , Dorado, J. , Torretta, J.P.** (2019), 'Scientific note: first global report of a bee nest built only with plastic', *Apidologie* (2019) 50:230–233.

Everaars, J., Strohbach, M. W., Gruber, B., Dormann, C.F. (2011), 'Microsite conditions dominate habitat selection of the red mason bee (Osmia bicornis, Hymenoptera: Megachilidae) in an urban environment: A case study from Leipzig, Germany', *Landscape and Urban Planning,* 103: 15–23.

Descriptions of *Anthophora* nests and of many other solitary bees are in **Chris O'Toole & Anthony Raw** (1991), *Bees of the World,* Blandford. The lining from the Dufour's gland secretion is tough enough for nests to survive for some time – they can be removed from cliffs, for example.

Else, G.R. and Edwards, M. (2018), op. cit., report fossilised *Anthophora* nest cells found in the Canary Islands dated to 25–30,000 years BP.

There are excellent photos of *Anthophora* nests in a cob wall on Christopher Wren's TrogTrog-Blog site: https://trogtrogblog.blogspot.com/2015/05/a-wall-of-bees.html and https://trogtrogblog.blogspot.com/2022/04/back-to-wall-of-bees.html.

Orr, M., Griswold, T., Pitts, J.P., Parker, F.D. (2016), 'A new bee species that excavates sandstone nests', *Current Biology,* 26 (17): R779–R793. **Michael Orr**'s website: https://michaelorr.weebly.com is also well worth a look.

Drawings of *A. plumipes* using her pygidium to tamp down tunnel walls taken from a video by **alrunen** on YouTube: https://www.youtube.com/watch?v=NAN2sUb1FPU (last accessed Jan 2019). A female sealing her nest is in a video by **John Walters**: https://www.youtube.com/watch?v=05VOGeCy9C0.

An article by **John Walters** in the BWARS newsletter of spring 2021 describes how to make small cob nest boxes for *Anthophora.* See video: https://www.youtube.com/watch?v=dAksl8Z6JaM – making a cob brick. Cob bricks can also be bought commercially: https://earthblocks.co.uk/about/

Pp.47–50: The drawings in the 'time-vine' are based on images and information in **Grimaldi, D. & Engel, M.S.** (2005), *Evolution of the Insects*, Cambridge University Press. **Danforth, B.N. & Poinar G.O.** (2011), 'Morphology, Classification and Antiquity of *Melit-tosphex burmensis* (Apoidea, Melittosphecidae) and implications for early bee evolution', *Journal of Palaeontology*, 85(5): 882–891.

Poinar, G.O. & Danforth, B.N. (2006), 'A Fossil Bee from Early Cretaceous Burmese Am-ber', *Science*, 314 (5799), p.614.

Poinar, George (2012), 'Desiomorphs in amber', *American Entomologist*, 58, 10.1093/ae/58.4.214.

Melittosphex burmensis is now considered a wasp and was removed from Apoidea in 2021: Rosa **B.B. & Melo G.** (2021), 'Apoid wasps (Hymenoptera: Apoidea) from mid-Cretaceous amber of northern Myanmar', *Cretaceous Research*, 122, 104770 (not open access).

Chambers, K.L., Poinar, G., Buckley, R. (2010), '*Tropidogyne*, a New Genus of Early Cre-taceous Eudicots (Angiospermae) from Burmese Amber', *Novon* 20: 23–29.

Crepet, W.L., Nixon, K.C., Gandolfo, M.A. (2004), 'Fossil Evidence and Phylogeny: The age of Major Angiosperm Clades based on Mesofossil and Macrofossil Evidence from Cre-taceous Deposits', *American Journal of Botany* 91(10): 1666–1682 (image of *Palaeoclusia*). **Engel, M.S.** (2000), 'A New Interpretation of the Oldest Fossil Bee (Hymenoptera: Apidae)' *American Museum Novitates*, no. 3296, American Museum of Natural History, 25/4/2000 (on *Cretotrigona*).

Michez, D., De Meulemeester, T., Rasmont, P., Nel, A., Patiny, S. (2009), 'New fossil ev-idence of the early diversification of bees: *Paleohabropoda oudardi* from the French Pale-ocene (Hymenoptera, Apidae, Anthophorinae)', *Zoologica Scripta*, 38(2), 171–181, March 2009.

Crepet, W., Nixon, K.C. (1998), 'Fossil Clusiaceae from the Late Cretaceous (Turonian) of New Jersey and implications regarding the history of bee pollination', *American Journal of Botany*, 85 (9): 1122–1133.

Michez, D., Vanderplanck, M., Engel, M.S. (2012), 'Fossil bees and their plant associ-ates', chapter 5 in *Evolution of Plant-Pollinator Relationships*, ed. S. Patiny, Cambridge Uni-versity Press. The tree in the centre of the four-page spread is based on the living Bur-mese tree *Agathis borneensis*, a relative of the ancient amber trees. The vine is the modern bat-leaved passion flower *Passiflora coriacea* (no particular connection to bees: I used to grow it). The bronze age bee pendant is from the Chrysolakkos funerary building at Ma-lia, northern Crete, dating from c.1850–1700 BC, the Middle Minoan period, described in **J. Lesley Fitton** (2002), Minoans, British Museum Press (Folio Society Edition, 2004). There is some debate as to whether these are actually bees: *Polistes* paper wasps or even *Megascolia*, the mammoth wasp have been suggested. I prefer honeybees, see: **Kitchell, K. F.** (1981), *Antiquity*, 55 (213): https://www.cambridge.org/core/journals/antiquity/article/mallia-wasp-pendant-reconsidered/12E7954326545F106642C3E277AADB52: honey was of course a key resource.

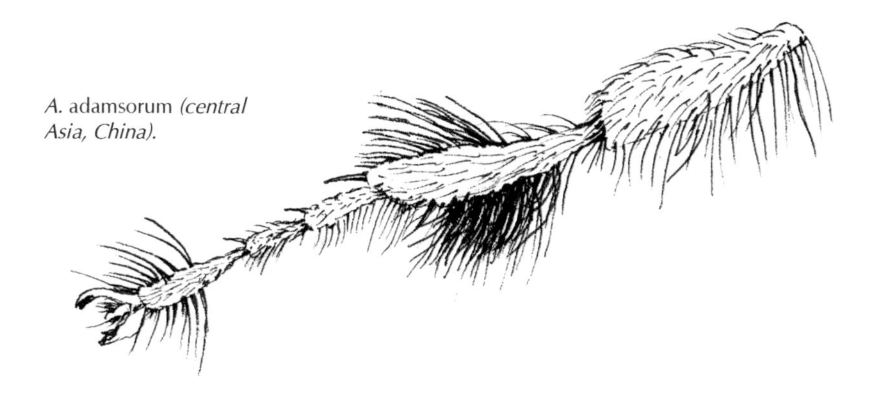

A. adamsorum (central Asia, China).

Pp.51–56: Grimaldi, D. and Engel, M.S. (2005), op. cit., discuss the origins of holometabolism. Interesting discussion also in **Bradley, T.J., Briscoe, A.D., Brady, S.G., et al.** (2009), 'Episodes in Insect Evolution', *Integrative and Comparative Biology,* 49 (5): 590–606.

See also **Gilbert, S.F.** (2000), *Metamorphosis: The Hormonal Reactivation of Development,* in *Developmental Biology,* Sinauer Associates (OUP), chapter accessible to view at: https://www.ncbi.nlm.nih.gov/books/NBK9986/; also: **Ian Stell** (2012), op. cit.

Dufour's gland secretions used as larval food in *Anthophora* species is reported in **Norden, B.B., Batra, S.W.T., Fales, H.M., Hefetz, A., Shaw, G.J.,** (1980), '*Anthophora* bees: unusual glycerides from maternal Dufour's glands serve as larval food and cell lining', *Science* 207(4435): 1095–1097 (not open access). Also discussed in **Danforth, Minckley and Neff**, *The Solitary Bees* op.cit.

Mark L. Winston (1987), op. cit.; **Ian Stell** (2012), op. cit.; **Snodgrass, R.E.** (1910, 1956), op. cit.

Baer, B., Eubel, H., Taylor, N.L., O'Toole, N., Millar, A.H. (2009), 'Insights into female sperm storage from the spermathecal fluid proteome of the honeybee *Apis mellifera*', *Genome Biology*. 10 (6): R67.

Pascini, T.V., Martins, G.F. (2017), 'The insect spermatheca: an overview', *Zoology,* 121: 56–71 (abstract only on free access).

Borgia, G., (1980), 'Evolution of Haplodiploidy: Models for Inbred and Outbred Systems', *Theoretical Population Biology* 17: 103–128.

Normark, B.B. (2003), 'The Evolution of Alternative Genetic systems in Insects', *Annual Review of Entomology.* 48: 397–423.

Normark, B.B., Ross, L. (2014), 'Genetic conflict, kin and the origins of novel genetic systems', *Philosophical Transactions of the Royal Society B,* 369: 20130364.

Kuijper, B., Pen, I. (2009), 'The evolution of haplodiploidy by male-killing endosymbionts: importance of population structure and endosymbiont mutualisms', *Journal of Evololutionary Biology,* 23: 40–52.

Ross, L., Shuker, D.M., Normark, B.B., Pen, I. (2012), 'The role of endosymbionts in the evolution of haploid-male genetic systems in scale insects (Coccoidea)', *Ecology and Evolution* 2 (5): 1071–1081.

Mason bees and parasites: **Seidelmann, K.** (2006), 'Open-cell parasitism shapes maternal investment patterns in the Red Mason bee *Osmia rufa*', *Behavioural Ecology* 17 (5): 839–848.

And in *Megachile* (abstract only): **Peterson, J.H., Roitberg, B.D.** (2006), 'Impacts of flight distance on sex ratio and resource allocation to offspring in the leafcutter bee, *Megachile rotundata*', *Behavioural Ecology and Sociobiology,* 59 (5): 589–596.

Pp.57–58: The Trellick Bee Tower (also pp.77-78 & 135) was sponsored by Community Service Volunteers to launch their Action Earth Campaign in 2010. It was designed and built by the author with the help of Martha Macdonald and Sarah Wilson and was erected in the coach park next to the real Trellick Tower in north London, for children from Middle Row Primary School to 'cut the ribbon'. It is a good mimic of the real tower. It was then given its permanent home in the Wildlife Garden at Roots and Shoots. It was redeveloped in 2020 to become the Bee Shard. The drawing of the hut in the woods is based on Thoreau's cabin at Walden Pond, from a drawing by his sister, Sophia. **Henry David Thoreau** was a 19th-century American environmentalist/writer/philosopher. The quotation is from *Walden, or, Life in the Woods*, (chapter 5, 'Solitude'), Boston, 1854.

Pp.59–64: Černá, K., Zemenová, M., Macháčková, L., Kolínová, Z., Straka, J. (2013), 'Neighbourhood Society: Nesting Dynamics, Usurpations and Social Behaviour in Solitary Bees', *PLOS ONE* 8 (9): e73806.

For more on eusocial insects (I am not including here the enormous literature on honeybee colonies) see the books by Edward Wilson, the (late) 'grandfather' of eusocial insects (and sociobiology). They include:

Bert Hölldobler and E.O. Wilson (2009), *The Super-organism: the Beauty, Elegance, and Strangeness of Insect Societies*, Norton; its predecessor, **Edward O. Wilson** (1971), *The Insect Societies*, Harvard (worth seeking second hand); and **Bert Hölldobler and E.O. Wilson** (1994), *Journey to the Ants: A story of scientific exploration*, Belknap, Harvard.

Also see **Wenseleers, T.** (2009), 'The Superorganism Revisited', *Bioscience* 59: 702–705, a review.

Thomas D. Seeley (1995), *The Wisdom of the Hive: the Social Physiology of Honey Bee Colonies*, Harvard University Press, and *Honeybee Democracy* (2010) Princeton University Press (both are wonderful books).

Hofstadter, D.R. (1979), *Gödel, Escher, Bach: An Eternal Golden Braid,* Basic Books.

Page, R.E. (2013), *The Spirit of the Hive: the Mechanisms of Social Evolution*, Harvard University Press.

Papers on eusociality

Wilson, E.O. and Hölldobler, B. (2005), 'Eusociality: Origin and Consequences', *PNAS* 102 (38): 13367–13371.

Wilson, E.O. (2008), 'One Giant Leap: How Insects Achieved Altruism and Colonial Life', *Bioscience* 58 (1): 17–25.

Nowak, M.A., Tarnita, C.E., Wilson, E.O. (2010), 'The Evolution of Eusociality', *Nature* 466 (7310): 1057–1062.

Cardinal, S., Danforth, B.N. (2011), 'The Antiquity and Evolutionary History of Social Behaviour in Bees', *PLOS ONE* 6 (6): e21086.

Danforth, B.N. (2002), 'Evolution of sociality in a primitively eusocial lineage of bees', *PNAS* 99 (1): 286–290.

Hunt, J.H. (2011), 'A conceptual model for the origin of worker behaviour and adaptation of eusociality', *Journal of Evololutionary Biology*, 25: 1–19.

Gibbs, J., Brady, S.G., Kanda, K., Danforth, B.N. (2012), 'Phylogeny of halictine bees supports a shared origin of eusociality for *Halictus* and *Lasioglossum* (Apoidea: Anthophila: Halictidae)', *Molecular Phylogenetics and Evolution* 65 (3): 926–939.

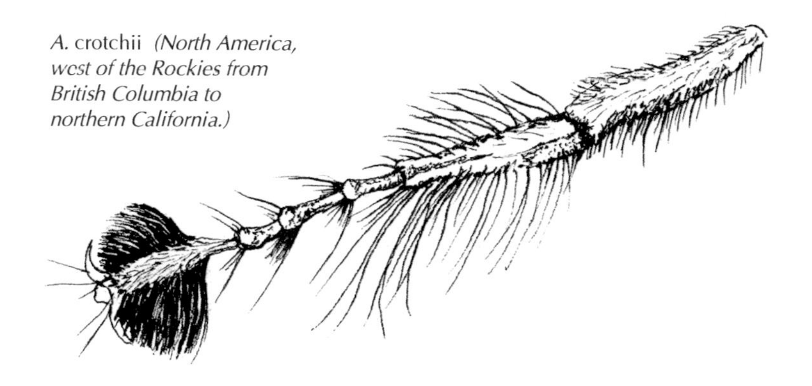

A. crotchii *(North America, west of the Rockies from British Columbia to northern California.)*

Johnstone, R. A., Cant, M. A., & Field, J. (2011), 'Sex-biased dispersal, haplodiploidy and the evolution of helping in social insects', *Proceedings Biological Sciences*, 279 (1729): 787–93.

Gardner, A., Alpedrinha, J., West, S.A. (2012), 'Haplodiploidy and the Evolution of Eusociality: Split Sex Ratios', *The American Naturalist*, Feb. 2012.

Alpedrinha, J., West, S.A., Gardner, A. (2013), 'Haplodiploidy and the Evolution of Eusociality: Worker Reproduction', *The American Naturalist*, Oct. 2013.

Lars Chittka (2022) discusses difficulties in the concept of the 'hive mind', suggesting that intelligence in the social insect resides, still, in the individual; there is no evidence, as yet, of intelligent behaviour by the hive (or swarm) as a whole unit (*The Mind of a Bee*, Princeton).

P.61: on *Halictus rubicundus*:

Soro, A., Field, J., Bridge, C., Cardinal, S.C., Paxton, R.J. (2010), 'Genetic differentiation across the social transition in a socially polymorphic sweat bee, *Halictus rubicundus*', *Molecular Ecology* 19 (16): 3351–3363 (abstract).

Field, J., Paxton, R., Soro, A., Craze, P., Bridge, C. (2012), 'Body size, demography and foraging in a socially plastic sweat bee: a common garden experiment', *Behavioral Ecology and Sociobiology* 66: 743–756.

Field, J., Paxton, R., Soro, A., Bridge, C. (2010), 'Cryptic Plasticity Underlies a Major Evolutionary Transition', *Current Biology* 20: 2028–2031.

Soucy, S.L., Danforth, B.N. (2002), 'Phylogeography of the socially polymorphic sweat bee *Halictus rubicundus* (Hymenoptera: Halictidae)', *Evolution* 56 (2): 330–341.

Brady, S.G., Sipes, S., Pearson, A., Danforth, B.N. (2006), 'Recent and simultaneous origins of eusociality in halictid bees', *Proceedings of the Royal Society B* 273: 1643–1649.

Keller, L. (2003), 'Behavioral Plasticity: Levels of Sociality in Bees', *Current Biology* 13: R644–645.

Chapuisat, M. (2010), 'Evolution: Plastic Sociality in a Sweat Bee', *Current Biology* 20 (22): R977–979.

Davison, P.J. (2016), 'Social Polymorphism and Social Behaviour in Sweat bees (Hymenoptera: Halictidae)', PhD thesis, University of Sussex.

Davison, P.J., Field, J. (2016), 'Social polymorphism in the sweat bee *Lasioglossum (Evylaeus) calceatum*', Insectes Sociaux, 63: 327–338.

Davison, P.J., Field, J. (2018), 'Limited social plasticity in the socially polymorphic sweat bee *Lasioglossum calceatum*', *Behavioral Ecology and Sociobiology* 72 (3): 56.

Séguret, A., Bernadou, A., Paxton, R.J. (2016), Facultative social insects can provide insights into the reversal of the longevity/fecundity trade-off across the eusocial insects', *Current Opinion in Insect Science* 16: 95–103.

Ulrich, Y., Perrin, N., Chapuisat, M. (2009), 'Flexible social organization and high incidence of drifting in the sweat bee, *Halictus scabiosae*', *Molecular Ecology* 18: 1791–1800.

Danforth, B.N., Conway, L., Ji, S. (2003), 'Phylogeny of eusocial *Lasioglossum* reveals multiple losses of eusociality within a primitively eusocial clade of bees (Hymenoptera: Halictidae)', *Systematic Biology*, 52 (1): 23–36.

Research on *H. rubicundus* continued at the Knepp Estate, Sussex:

Boulton, R.A., Field, J. (2022), 'Sensory plasticity in a socially plastic bee' bioRxiv 2022.01.29.478030. doi: https://doi.org/10.1101/2022.01.29.478030 or https://academic.oup.com/jeb/article/35/9/1218/7317852 – see 'Bees Moving North', pp.125-126.

For references on the 'social brain hypothesis' and how it might apply in insects, see notes for **p.14** above.

Pp.65–68: The drawings of the evolution of the sting are based on diagrams in **Grimaldi, D., Engel M.S.** (2005), *Evolution of the Insects*, Cambridge University Press, and in **Goodman, L.** (2003), *Form and Function in the Honey Bee,* IBRA. Drawing of honey bee stinging based on photo by Waugsberg (Wikimedia Commons). The drawings of wasps, bees and nests are based upon combinations of observation, photographs and images on the web.

Pp.69–70: Dates for flying seasons of species shown compiled from the BWARS website and **Falk, S.** (2015) op. cit. Records of new species and 'unseasonal' flights are often reported (**Adrian Knowles** for BWARS) in *British Wildlife*, e.g. vol. 30, no.3, Feb. 2019.

Pp.71–72: Drawing of female based on photos of my own and some on the web including: https://www.flickr.com/photos/elbowes/7118664629/in/set-72157629153643586/ and https://www.flickr.com/photos/33883829@N05/7166295047/in/set-72157629496536452; and of the male based on photos on BWARS by **Nigel Jones** and by **Steven Falk** on https://www.flickr.com/photos/63075200@N07/26393824843/in/photostream/. Nest architecture, **Christopher O'Toole** (1999), *Bees of the World*, op. cit. p.41–2.

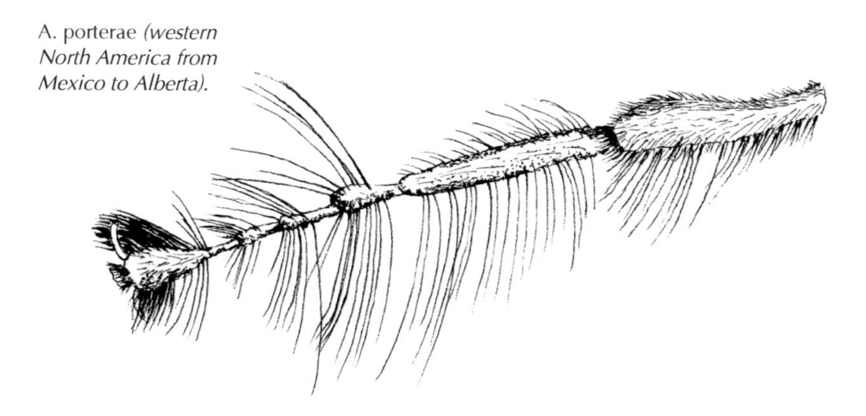
A. porterae *(western North America from Mexico to Alberta).*

Pp.73–74: For an investigation into the importance of habitat components in solitary bee breeding see: **Everaars, J., Strohbach, M.W., Gruber, B., Dormann, C.F.** (2011), 'Microsite conditions dominate habitat selection of the red mason bee (*Osmia bicornis*, Hymenoptera: Megachilidae) in an urban environment: a case study from Leipzig, Germany', *Landscape and Urban Planning* 103: 15–23.

Also very useful here: **Kirk and Howes** (2012), *Plants for Bees,* IBRA.

Jan Miller-Klein (2010), *Gardening for Bees, Butterflies and other beneficial insects*, Saith Ffynnon Books.

The **Urban Pollinators Project** http://www.bristol.ac.uk/biology/research/ecological/community/pollinators is just one website amongst many that are now discussing key plants for a healthy population of bees.

See also **Marc Carlton**, *The Pollinator Garden* website: http://www.foxleas.com.

Wildlife Gardening Forum (WLGF): http://wlgf.org/plants_planting_intro.html#gsc.tab=0, and for plants to avoid: http://wlgf.org/plants_avoid.html#gsc.tab=0.

For plants: https://www.7wells.co.uk; http://www.rosybee.com (also giving advice for gardeners and research on plants for bees).

On *Anthidium* and *Stachys*: **Strange, J.P., Koch, J.B., Gonzalez, V.H., Nemelka, L., Griswold, T.** (2011), 'Global invasion by *Anthidium manicatum* (Linnaeus) (Hymenoptera: Megachilidae): assessing potential distribution in North America and beyond', *Biological Invasions* 13 (9): 2115 –2133.

Griswold, T., Gonzalez, V.H., Ikerd, H. (2014), 'AnthWest, occurrence records for wool carder bees of the genus Anthidium (Hymenoptera, Megachilidae, Anthidiini) in the Western Hemisphere', *ZooKeys* 408: 31–49.

Status of *Stachys byzantina* in Britain: https://www.brc.ac.uk/plantatlas/plant/stachys-byzantina.

Pp.75–76: Seidelmann, K. (1999), 'The Race for Females: The Mating System of the Red Mason Bee, *Osmia rufa* (L.) (Hymenoptera: Megachilidae)', *Journal of Insect Behavior* 12 (1): 13–25.

Conrad, T., Paxton, R.J., Barth, F.G., Francke, W., Ayasse, M. (2010), 'Female choice in the red mason bee, *Osmia rufa* (L.) (Megachilidae)', *Journal of Experimental Biology* 213: 4065–4073.

Kierat, J., Szentgyörgyi, H., Woyciechowski, M. (2017), 'Orientation inside linear nests by male and female *Osmia bicornis* (Megachilidae)', *Journal of Insect Science* 17 (2): 40.

Szentgyörgyi, H., Woyciechowski, M. (2012), 'Cocoon orientation in the nests of red mason bees (*Osmia bicornis*) is affected by cocoon size and available space', *Apidologie* 44 (3): 334–341.

Pp.77–78: There is now a vast library of examples of solitary bee nest boxes/'trap nests'/bee hotels on the web – there even seems to be a competitive element to both size and variety of design. It is possible to come across bee hotel structures in almost any park, community garden or public open space in London and elsewhere. Many of those I have come across are empty of nests, mainly due to being placed on tree trunks and/or in heavy shade or in spaces where other components – forage/nest materials are absent (see the Leipzig study referenced above, p.73) – or in some cases where there is a big and very competitive honeybee apiary nearby. There is a maximum size, it seems to me, beyond which there is little point in trying to artificially generate an unstable, high population density of any particular bee species: the high number of honeybee apiaries and beekeepers in London is potentially doing significant damage to bee species diversity. Keep your nest boxes relatively simple and aim for diversity.

See also chapter 14 on bee projects in **Christopher O'Toole,** (2013), *Bees: A Natural History* and in his essay for *Plants for Bees*, **Kirk and Howes**, (2012), IBRA.

Also in **Jan Miller-Klein,** (2010) op. cit. The student work shown is by Nora Kurayshi and Pavlina Vackova. Martha's artwork: http://www.marthamacdonald.co.uk.

See also reviews of papers in *Conservation Evidence*, e.g. https://www.conservationevidence.com/actions/47 and **Dicks, L.V., Showler, D.A. & Sutherland, W.J.** (2010) *Bee conservation: evidence for the effects of interventions*, Pelagic Publishing, Exeter (see https://www.conservationevidence.com/synopsis/pdf/7).

Lots of discussion and guidance on making solitary bee nest boxes is on the **Wildlife Gardening Forum** website (http://www.wlgf.org/ht_bee_hotel.pdf; http://www.wlgf.org/bee_hotels_issues.pdf; http://www.wlgf.org/ht_improve_pollinators.pdf) and **Dave Goulson**'s YouTube channel: https://www.youtube.com/@davegoulson6831/videos.

Pp.79–80: The behaviour of territorial male *Anthidium manicatum* around suitable flower patches, as described by **Christopher O'Toole** in *Bees: A Natural History*, sounds very much like the behaviour of many territorial mammals. The successful male, who secures a patch of productive flowers and reserves it for his own and his females' use, attracts more females seeking a mating. The following papers (abstracts only) investigate *Anthidium* behaviour:

Wirtz, P., Szabados, M., Pethig, H., Plant, J. (1988), 'An Extreme Case of Interspecific Territoriality: Male Anthidium manicatum (Hymenoptera: Megachilidae) Wound and Kill Intruders', *Ethology* 78: 159–167.

Starks, P.T., Reeve, H.K. (1999), 'Condition-based alternative reproductive tactics in the wool-carder bee, Anthidium manicatum', *Ethology, Ecology & Evolution* 11(1): 71–75.

Severinghaus, L.L., Kurtak, B.H., Eickwort, G.C. (1981), 'The reproductive behaviour of *Anthidium manicatum* (Hymenoptera: Megachilidae) and the significance of size for territorial males', Behavioural Ecology and Sociobiology 9 (1): 51–58.

Lampert, K.P., Pasternak, V., Brand, P., Tollrian R., Leese, F., Eltz, T. (2014), '"Late" male sperm precedence in polyandrous wool-carder bees and the evolution of male resource defence in Hymenoptera', *Animal Behaviour* 90: 211–217.

There is a video of females nest-building on the BWARS website or on YouTube: https://www.youtube.com/watch?time_continue=22&v=6utPWbIBdKw.

In 2021 there was an intriguing new record for Britain of *Anthidium septemspinosum*, a striking species from the near continent, in Hereford (see wildlife reports in British Wildlife, 33 (3), p.15, December 2021 and referred to here on p.119).

Spot the bee at Waterloo East!

Pp.81–82: An interesting study of leaf preference in Canada is here:

J. Scott MacIvor (2016), 'DNA barcoding to identify leaf preference of leafcutting bees', *Royal Society Open Science*, 3: 150623.

Pp.83–84: Litman, J.R., Danforth, B.N., Eardley, C.D., Praz, C.J. (2011), 'Why do leafcutter bees cut leaves? New insights into the early evolution of bees', *Proceedings of the Royal Society B*, 278: 3593–3600.

An interesting fossil representing a leaf used by leaf-cutter bees is described here: **Sarzetti, L.C., Labandeira C.C., Genise, J.F.** (2008), 'A leafcutter bee trace fossil from the Middle Eocene of Patagonia, Argentina, and a review of Megachilid (Hymenoptera) ichnology', *Palaeontology*, 51 (4): 933–941.

Wittmann, D., Blochtein, B. (1995), 'Why males of leafcutter bees hold the females' antennae with their front legs during mating', *Apidologie,* 26 (3): 181–196.

Guédot, C., Buckner, J.S., Hage, M.M., Bosch, J., Kemp, W.P., Pitts-Singer, T.L. (2013), 'Nest marking behaviour and chemical composition of olfactory cues involved in nest recognition in *Megachile rotundata*', *Environmental Entomology,* 42 (4): 779–789.

Pp.85–86: The London Wildlife Trust's Centre for Wildlife Gardening has been going since 1989: https://www.wildlondon.org.uk/reserves/centre-for-wildlife-gardening.

RSPB Flatford Wildlife Garden: https://www.rspb.org.uk/reserves-and-events/reserves-a-z/flatford-wildlife-garden/; photo, Amy Ward, RSPB. Three Hagges Woodmeadow image courtesy of The Woodmeadow Trust/Dave Raffaelli. In August 2023 the meadow became a Plantlife reserve: https://www.plantlife.org.uk/our-work/three-hagges-woodmeadow-nature-reserve/.

Pp.87–90: Falk, S (2015), op. cit. **George R. Else and Mike Edwards** (2018), op. cit. **Magnacca, K.N., Danforth, B.N.** (2006), 'Evolution and biogeography of native Hawaiian *Hylaeus* bees (Hymenoptera: Colletidae)', *Cladistics* 22 (5): 393–411.

Magnacca, K.N., Danforth, B.N. (2007), 'Low nuclear DNA variation supports a recent origin of Hawaiian *Hylaeus bees* (Hymenoptera: Colletidae)', *Molecular Phylogenetics and Evolution* 43 (3): 908–915.

Magnacca, K. (2007), 'Conservation Status of the Endemic Bees of Hawai'i, *Hylaeus* (*Nesopropis*) (Hymenoptera: Colletidae)', *Pacific Science*, 61 (2): 173–190.

Wianecki, S. (2015), 'Local Buzz', *Maui Magazine,* Nov–Dec, 2015.

Kayaalp, P., Schwarz, M.P., Stevens, M.I. (2013), 'Rapid diversification in Australia and two dispersals out of Australia in the globally distributed bee genus, *Hylaeus* (Colletidae: Hylaeinae)', *Molecular Phylogenetics and Evolution*, 66 (3): 668–687.

Cardinal, S., Danforth, B.N. (2013), 'Bees diversified in the age of the eudicots', *Proceedings of the Royal Society B*, 280: 20122686.

Hedtke, S.M., Patiny, S., Danforth, B.N. (2013), 'The bee tree of life: a supermatrix approach to apoid phylogeny and biogeography', *BMC Evolutionary Biology,* 13: 138.

McLendon, R., (2016), 'U.S. declares bees endangered for first time', *Mother Nature Network*, 17 August 2018.

Almeida, E.A.B (2008), 'Colletidae nesting biology (Hymenoptera: Apoidea'), *Apidologie*, 39: 16–29.

Almeida, E.A.B., Danforth, B.N. (2009), 'Phylogeny of colletid bees (Hymenoptera: Colletidae) inferred from four nuclear genes', *Molecular Phylogenetics and Evolution,* 50: 290–309.

Drawing of glossa of *Colletes daviesanus* from the male caught by the spider.

See also **Hefetz, A.** (1987) 'The role of Dufour's gland secretions in bees', *Physiological Entomology,* 12, 243–253.

Müller, A., Weibel, U. (2020), 'A scientific note on an unusual hibernating stage in a late-flying European bee species', *Apidologie,* 51: 436–438 https://doi.org/10.1007/s13592-019-00730-8.

Hennessy, G., Uthoff, C., Abbas, S., Quaradeghini, S., Stokes, E., Goulson, D., Ratnieks, F. (2021), 'Phenology of the specialist bee *Colletes hederae* and its dependence on *Hedera helix* L. in comparison to a generalist, *Apis mellifera*', *Arthropod-Plant Interactions*, 15:183–195: discusses monolecty, the dependence on ivy and potential competition for forage with other pollinating insects.

Carreck NL, Andernach J, Ariss A, Dowd H, Gant A, Garbuzov M, Hennessy G, Nash L,Stagg A, Ratnieks FLW (2023) Distribution and abundance of the ivy bee, *Colletes hederae* Schmidt & Westrich, 1993, in Sussex, southern England. *BioInvasions Records* **12**(3): 681–697, https://doi.org/10.3391/bir.2023.12.3.06

Pp.91–96: Falk, S. (2015), op. cit.

George R. Else and Mike Edwards (2018), op. cit.

See also references given for *Halictus rubicundus* **p.61**.

The ID of the female *Sphecodes gibbus* is based on several photographs at the same site and from different years but cannot be guaranteed – she could be *S. monilicornis*. Both species use *H. rubicundus* as host.

For more on *Heriades truncorum* see **p.82**.

Pp.97–98: Brooks, R.W. (1988), 'Systematics and Phylogeny of the Anthophorine Bees (Hymenoptera: Anthophoridae; Anthophorinae)', *University of Kansas Science Bulletin*, 53 (9): 436–575. Brooks makes the following comments on the origins and biogeography of Anthophorine bees (p.524): 'The greatest diversity of Anthophorine species and subgenera occurs in the Mediterranean regions, which is the tribe's possible place of origin. Of the 14 subgenera of *Anthophora*, eight occur in the New World, of which seven are also Palearctic and only one limited to North America. It is therefore probable that the genus originated in the Old World. The genus *Anthophora* is widespread but conspicuously absent from the Indomalaysian–Australian area. Its absence there is probably due to its being ill-adapted to moist tropical climates through which it would have had to pass, given the present continental arrangement, to reach the temperate parts of Australia. Where it has penetrated the tropics, as in Africa and South America, *Anthophora* is found in montane or xeric habitats. *Anthophora* is apparently a good island colonizer; it is found in the Cape Verde Islands, Canary Islands, numerous Mediterranean islands as well as the Greater and Lesser Antilles.'

Pp.99–100: Benton, T. (2017), *Solitary Bees*, Naturalists' Handbook 33, Pelagic Publishing.

Pp.101–102: David Notton, formerly of the Natural History Museum, London, (now at National Museums of Scotland) has undertaken a thorough survey of the bees on Greenwich Peninsula: **Notton, D.** (2018), 'Bees, wasps, flowers and other biological records from Greenwich Peninsula Ecology Park, Southern Park and the Greenwich Peninsula: records made 2016–2017', *ResearchGate*. A spreadsheet of species can be downloaded. Also for Blackheath: https://www.researchgate.net/publication/330684522_Bees_wasps_and_ants_aculeate_Hymenop-tera_with_flower_visit_data_from_Blackheath_south_east_London_UK_records_made_2009-2018

Drawing of *Dasypoda* larva and nest based on online images by **M. Gosselin.**

Drawing of female working from videos on YouTube: https://www.youtube.com/watch?v=oKDJQFYoac8 by **John Walters**, and https://www.youtube.com/watch?v=9IMd1fZc3eg by **Roy Kleukers.**

Pp.103–104: The remarkable anatomy of *Gasteruption* and other parasitoid wasps such as the ichneumon, *Ephialtes manifestator,* shown on p. 66 deserves celebration and a book of its own... for more on *Gasteruption* see: http://tolweb.org/Gasteruption/25832.

The first pair of Irish Hairy-footed flower bees, male this page, female opposite, in Dublin, March 2022. Photos by Martin Molloy.

Pp.105–110: Nick Davies (2015), *Cuckoo: Cheating by Nature*, Bloomsbury.

Canestrari, D., Bolopo, D., Turlings, T.C.J., Röder, G., Moarcos, J.M., Baglione, V. (2014), 'From Parasitism to Mutualism: Unexpected Interactions Between a Cuckoo and Its Host', *Science*, 343 (6177): 1350–1352.

Soler, M., de Neve, L., Roldán, M., Pérez-Contreras, T., Soler, J.J. (2017), 'Great spotted cuckoo nestlings have no antipredatory effect on magpie or carrion crow nests in southern Spain', *PLOS ONE* 12 (4): e0173080.

An example of weird parasitism: **Weinersmith, K.L., Liu, S.M., Forbes, A.A., Egan, S.P.** (2017), 'Tales from the crypt: a parasitoid manipulates the behaviour of its parasite host', *Proceedings of the Royal Society B*, 284: 20162365.

Paxton, R.G., Pohl, H. (1999), 'The tawny mining bee, *Andrena fulva* (Müller) (Hymenoptera, Andreninae), at a South Wales field site and its associated organisms: Hymenoptera, Diptera, Nematoda and Strepsiptera', *British Journal of Entomology and Natural History*, 12: 56–67.

Baldock, D., (2010), *Wasps of Surrey*, Surrey Wildlife Trust.

Great images on the web of cuckoo wasps e.g.: https://www.chrysis.net.

Also by **Guillermo Booth** on https://www.biodiversidadvirtual.org/insectarium/Chrysura-radians-cat36931.html.

BWARS website;

Morgan, D., (1984), *Cuckoo-Wasps: Hymenoptera, Chrysididae, Handbooks for the Identification of British Insects 6* (5), Royal Entomological Society of London (online version, 2013).

Paukkunen, J., Berg, A., Soon, V., Ødegaard, F., Rosa, P. (2015), 'An illustrated key to the cuckoo wasps (Hymenoptera, Chrysididae) of the Nordic and Baltic countries, with description of a new species', *ZooKeys* 548: 1–116.

Pp.105–110 contd.: Cardinal, S., Straka, J., Danforth, B.N. (2010), 'Comprehensive phylogeny of apid bees reveals the evolutionary origins and antiquity of cleptoparasitism', *PNAS* 107 (37): 16207–16211.

Litman, J.R., Praz, C.J., Danforth, B.N., Griswold, T.L., Cardinal, S. (2013), 'Origins, evolution and diversification of cleptoparasitic lineages in long-tongued bees', *Evolution, 67–10: 2982–2998*

Fifteen new species in the cuckoo bee genus *Epeolus* of North America are reported in **Onuferko, T.M.** (2018), 'A revision of the cleptoparasitic bee genus *Epeolus*, Latreille, for Nearctic species, north of Mexico (Hymenoptera, Apidae)', *ZooKeys* 755: 1–185.

A detailed review of the complexities of brood parasitism and the number of times it has emerged in bees is in **Danforth, Minckley and Neff** (2019), op. cit.

On cuckoo bee larvae: **O'Toole and Raw** (1991), op. cit.

Stephen, W.P., Bohart, G.E., Torchio, P.F., (1969), *The Biology and External Morphology of Bees*, Oregon State University.

Michener, C.D. (2007), The bees of the World, 2nd ed., Johns Hopkins University Press.

On *Coelioxys*: https://www.amentsoc.org/insects/fact-files/orders/coelioxys-key-high-res.pdf.

On *Nomada*: **Tengö, J., Bergström, G.** (1977), 'Cleptoparasitism and odor mimetism in bees: do *Nomada* males imitate the odor of *Andrena* females?' *Science,* 196 (4294): 1117–1119.

Schindler, M., Hofmann, M.M., Wittmann, D., Renner, S.S. (2018), 'Courtship behaviour in the genus *Nomada* – antennal grabbing and possible transfer of male secretions', *Journal of Hymenoptera Research* 65: 47–59. They show the remarkable entwining of antennae during mating to transfer the pheromone.

George R. Else and Mike Edwards (2018), op. cit.

Pp.111–114: On *Stylops*: **Straka, J, Rezkova, K., Batelka,J., Kratochvíl, L.,** (2011), 'Early nest emergence of females parasitised by *Strepsiptera* in protandrous bees (Hymenoptera Andrenidae)', *Ethology, Ecology & Evolution,* 23: 97–109: http://www.aculeataresearch.com/attachments/article/56/Stylops_Andrena_EEE.pdf; also http://sea-entomologia.org/IDE@/revista_62B.pdf; https://www.royensoc.co.uk/understanding-insects/classification-of-insects/strepsiptera/; http://www.nhm.ac.uk/discover/which-parasite-has-the-weirdest-way-of-life.html. Video of male emerging here: https://www.youtube.com/watch?v=L7gFya9hXQU;

Crab spiders: **Huey, S., Nieh, J.C.** (2017), 'Foraging at a safe distance: crab spider effects on pollinators', *Ecological Entomology*, 42 (4).

Pp.115–124: The various ramifications of climate change and its impacts on British wildlife are discussed in **Trevor Beebee**'s book, *Climate Change and British Wildlife* (2018), British Wildlife Collection, no. 6, Bloomsbury. He discusses the records for white deadnettle *Lamium album*, described by the Fitters in Oxfordshire (**Fitter, A.H. & Fitter, R.S.R.** (2002), 'Rapid changes in flowering time in British plants', *Science*, 296: 1689–1691).

Citizen science accuracy: **Falk, S, Foster, G, Comont, R, Conroy, J, Bostock, H, Salisbury, A, Kilbey, D, Bennett, J & Smith, B.** (2019, 'Evaluating the ability of citizen scientists to identify bumblebee (*Bombus*) species' *PLOS ONE*, vol. 14, no. 6, e0218614.

Colletes hederae crossing water: **Vivian Russell** notes the following in her record for Grange-over-Sands, Cumbria, submitted to the NBN Atlas: 'Dispersal seems to be across the water, not along the coast eg Heysham Cliffs to Jenny Brown's Point and from Jenny Brown's Point to Grange-over-Sands, the last two in very close proximity to the coast, just metres away. Two locations were wild, the third suburban.' A shorter journey, admittedly, than over the Irish Sea!

Anthophora plumipes/Anthidium manicatum records, **Ben Hargreaves**, Lancashire Wildlife Trust, pers. comm. 03/22: '*Anthophora plumipes* and *Anthidium manicatum* are a bit of an oddity re: N.W. records...the general conception is that it is a recent arrival, though whether this is a re-colonisation may be a possibility...there is a 1943 specimen of *A. plumipes* at Tullie House Museum in Cumbria (from Carlisle – ironically a town it has re-appeared in more recently) and A.E. Wright reported *Anthidium* as common in his garden for a number of years in the early 1940s. Gardner recorded the latter (pre-1900) in some of the Cheshire valleys/N. Wales, whilst *A. plumipes* was stated to be widespread and often locally abundant in the Lancs/Cheshire district (same author – pre-1900), with a record of the associated and distinctive *Sitaris muralis* in the Bollin Valley.

No doubt that *A. plumipes* disappeared until the late 2000s in the northwest – there are no specimens in Liverpool, Manchester or Tullie House (nor written accounts) and it seems inconceivable that such a distinctive species would not have been collected in the decades from 1900 until present.'

The new Irish record for 2022: https://biodiversityireland.ie/hairy-footed-flower-bee-spotted-in-ireland-for-the-first-time/ – a pair seen in a community gardening project in Harold's Cross, Dublin. It was seen there again in March and April 2023 after breeding successfully. Many thanks to Úna Fitzpatrick for keeping me updated.

George R. Else and Mike Edwards, (2018) *Handbook of the Bees of the British Isles*, The Ray Society.

Pp.115–124 contd.: Wool carder bee: so striking it is the earliest solitary bee species named and recorded by a naturalist: **Gilbert White,** *A Natural History of Selborne,* 1788:

1772: July 11, 1772 – *Drought has continued five weeks this day. Watered the rasp and annuals well.*

* There is a sort of wild bee frequenting the garden-campion for the sake of its tomentum, which probably it turns to some purpose in the business of nidification. It is very pleasant to see with what address it strips off the pubes, running from the top to the bottom of a branch, & shaving it bare with all the dexterity of a hoop-shaver. When it has got a vast bundle, almost as large as itself, it flies away, holding it secure between its chin and its forelegs.

Anthidium manicatum global distribution discussed in: **Strange, J.P., Koch, J.B., Gonzalez, V.H., Nemelka, L., Griswold, T.** (2011), 'Global invasion by *Anthidium manicatum* (Linnaeus) (Hymenoptera: Megacilidae): assessing potential distribution in North America and beyond', *Biological Invasions,* 13 (9): 2115–2133 (see also p.73).

Records in Derbyshire and the Midlands: **Kieron Huston**, Derbyshire Biological Records Centre, pers.comm. April 2022.

Anthidium manicatum and *Anthophora plumipes* in flight: https://www.flickr.com/photos/99927961@N06/40100543185 & http://www.insektenflug.de (**Rolf Nagel**).

Anthidium septemspinosum drawing derived from: http://www.michel-ehrhardt.fr/album-photos/mes-amis/hymenopteres/megachilidae/.

Xylocopa violacea discussion: https://www.bwars.com/content/xylocopa-britain.

Also in: **David Baldock** (2008), *Bees of Surrey*, Surrey Wildlife Trust, and **Adrian Knowles,** *British Wildlife*, 31 (5) June 2020: '*Xylocopa* sp. records from gardens in Ashton-on-Ribble, near Preston in 2015 and south Liverpool (2017, 2018) – both almost certainly imports from nearby timber merchants / warehouses / docks.' **Ben Hargreaves**, pers. comm., 2022.

London records for *Xylocopa*: 14 July 2003, Guildford (male); June 2004 Reigate Heath; 9 September 2006, Nunhead, Peckham. Personal London records: July 2005 brief sighting, Roots & Shoots; 11 May 2006, Roots & Shoots, good sighting of female investigating nest holes; 15 May 2006, another sighting; July 2016, seen nearby to Roots & Shoots and 28 September 2016, female on bindweed at Roots & Shoots. Records from near Ipswich and in East Sussex in 2022 have been followed by an early-emerged male in Alderminster, Warwickshire in April 2023: described in a discussion of northwards movement in several species (including *Anthophora quadrimaculata*) by **Adrian Knowles** in *British Wildlife*, April & October 2023.

Two interesting *Xylocopa* videos: https://www.youtube.com/watch?v=xXokYWPzhGk

https://www.youtube.com/watch?v=w35I7lz2kk4.

Osmia cornuta: see also notes for Pp.5-6. **Steven Falk** comments on flickr: https://www.flickr.com/photos/63075200@N07/sets/72157691453175962/.

Red mason bee *Osmia bicornis* receiving assistance to get into Ireland – **Úna Fitzpatrick,** pers. obs. 2022.

O. cornuta as pollinator examples: https://pollinature.net/wp-content/uploads/2020/02/Osmia-cornuta-as-a-pollinator-of-pear-fruit-and-seed-set.pdf; http://www.bulletinofinsectology.org/pdfarticles/vol60-2007-077-082maccagnani.pdf.

O. cornuta Cheshire record near Macclesfield on iRecord and NBN Atlas, confirmed by Matt Smith. Luc Verhelst.

Heriades truncorum: **David Baldock,** (2008) *Bees of Surrey*, Surrey Wildlife Trust.

Else, G.R. and Edwards, M. (2018), *Handbook* op. cit.

'Even small shifts in ecology…': **Trevor Beebee** (2018) op. cit. discusses changes in preferred larval food plants of Comma *Nymphalis c-album* and Brown Argus *Aricia agestis* butterflies, and change in the habitat niche occupied by Speckled wood *Pararge aegiria* with climate change likely spurring change in the latter two species.

Mountain *Osmia* species: **Else, G.R. and Edwards, M.** (2018), *Handbook* op. cit.

Colletes floralis:

http://www.gnhs.org.uk/machair/colletes_floralis.pdf

https://gnhs.org.uk/machair/colletes.pdf

http://www.habitas.org.uk/priority/species.asp?item=9599

https://www.gbif.org/species/1348742

https://pubmed.ncbi.nlm.nih.gov/21040051/

https://irishnaturalist.com/bees/northern-colletes-bee-colletes-floralis/

https://www.buglife.org.uk/get-involved/near-me/buglife-northern-ireland/northern-ireland-threatened-bee-report/

https://www.biodiversityireland.ie/wordpress/wp-content/uploads/Simple-guide-to-solitary-bees-in-Ireland-part-2-2016.pdf

Bumblebees:

Article by **Richard Comont** of Bumblebee Conservation Trust on climate change/bumblebees: https://www.bumblebeeconservation.org/what-does-a-changing-climate-mean-for-the-uks-bumblebees/

Also:

https://www.bumblebeeconservation.org/ginger-yellow-bumblebees/great-yellow-bumblebee/

Colletes daviesanus down the allotment!

Pp.115–124: Data providers for maps and species accounts:

Irish records from the **National Biodiversity Data Centre** in Ireland: https://biodiversityireland.ie; https://maps.biodiversityireland.ie. There are more *Xylocopa* records from Ireland still to be confirmed. Further comments on the Irish records from **Úna Fitzpatrick**.

Other records:

Adrian Knowles, *British Wildlife*, vol. 30 no. 3, February 2019 (autumn records); vol. 31 no. 5, June 2020 (*Xylocopa* records); vol. 32 no. 5 April 2021 (autumn records); vol. 33 no. 3, Dec 2021 (*Anthidium septemspinosum* confirmed; *Colletes hederae* reported from near Dunbar).

Bees, Wasps & Ants Recording Society: https://www.bwars.com/content/bwars-data-download (downloaded March 2022).

Biological Records Centre (downloaded March 2022). Bee, wasp and ant (Hymenoptera: Aculeata) records verified via iRecord.

South East Wales Biodiversity Records Centre (SEWBRec). Bees, wasps and ants (southeast Wales; downloaded March 2022). Occurrence dataset on the NBN Atlas.

Bumblebee Conservation Trust BeeWalk bumblebee distribution records for Great Britain 2008–2019. Data collected on behalf of the Bumblebee Conservation Trust. Occurrence dataset on the NBN Atlas.

Bristol Regional Environmental Records Centre (downloaded March 2022). The Centre's species records from all years at full resolution excluding notable species within the last ten years. Occurrence dataset on the NBN Atlas.

Shropshire Ecological Data Network (downloaded March 2022). Shropshire Ecological Data Network database. Occurrence dataset on the NBN Atlas.

National Trust (downloaded March 2022). National Trust Species Records. Occurrence dataset on the NBN Atlas.

Suffolk Biodiversity Information Service (SBIS) (downloaded March 2022). Occurrence dataset on the NBN Atlas dataset.

Biodiversity Information Service (BIS) for Powys & Brecon Beacons National Park (downloaded March 2022). Miscellaneous records held by BIS. Occurrence dataset on the NBN Atlas

(downloaded March 2022). Montgomeryshire Wildlife Trust records held by BIS. Occurrence dataset on the NBN Atlas.

Natural Resources Wales (downloaded March 2022). Welsh Invertebrate Database. Occurrence dataset accessed through the NBN Atlas.

Malcolm Storey (March 2022). http://www.bioimages.org.uk/. Malcolm Storey personal records and images. Occurrence dataset on the NBN Atlas.

Cofnod – North Wales Environmental Information Service. Miscellaneous records provided on the Cofnod database, Cofnod (2019), on behalf of the recorders of those records, whose rights are recognised.

Nottinghamshire Biological and Geological Records Centre (downloaded March 2022). Trevor and Dilys Pendleton Eakring and Nottinghamshire Invertebrate Records. Occurrence dataset on the NBN Atlas. UK abstract from Nottingham City Museums & Galleries (NCMG) Insect Collection Baseline database. NCMG (downloaded March 2022). Occurrence dataset on the NBN Atlas.

Staffordshire Ecological Record SER Species-based Surveys. Including data from Defra Family Organisations supplied to Staffordshire Ecological Record.

Royal Horticultural Society Records from the RHS insect reference collection (downloaded February 2022). RHS monitoring of native and naturalised plants and animals at its gardens and surrounding areas. Occurrence dataset on the NBN Atlas.

Natural England (downloaded March 2022) Invertebrate Common Standards Monitoring and ISIS Test Data. Occurrence dataset on the NBN Atlas.

Invertebrate Site Register – England (1738–2005). Occurrence dataset on the NBN Atlas.

Natural England iRecord Surveys (downloaded March 2022).

Merseyside BioBank (2020): www.merseysidebiobank.org.uk. Merseyside BioBank Recording Network (downloaded March 2022).

Leicestershire and Rutland Environmental Records Centre (downloaded March 2022). Leicestershire and Rutland Environmental Records Centre records 2015–2019. Occurrence dataset on the NBN Atlas. All taxa records for Leicestershire and Rutland NatureSpot.

Norfolk Biodiversity Information Service (NBIS) Records to December 2016: biological records provided by NBIS on behalf of the recorders of those records, whose rights are recognised (downloaded March 2022). Occurrence dataset on the NBN Atlas.

Derbyshire Biological Records Centre (DBRC); (downloaded March 2022). DBRC Hymenoptera species records 1975–2015. Occurrence dataset on the NBN Atlas.

RECORD Aculeate Hymenoptera Data Biological records for Cheshire provided by RECORD, on behalf of the recorders of those records, whose rights are recognised (downloaded March 2022). Occurrence dataset on the NBN Atlas.

Cumbria Biodiversity Data Centre, Tullie House Museum Natural History Collections (downloaded March 2022). Occurrence dataset on the NBN Atlas.

Rotherham Biological Records Centre (download March 2022). Non-sensitive records from all taxonomic groups. Occurrence dataset on the NBN Atlas.

Highland Biological Recording Group (HBRG); (downloaded March 2022). HBRG Insects Dataset. Occurrence dataset accessed through the NBN Atlas.

Argyll Biological Records Centre (downloaded March 2022). Argyll Biological Records Dataset. Occurrence dataset on the NBN Atlas.

Lancashire Environment Record Network (LERN); (download March 2022). LERN Records; Occurrence dataset on the NBN Atlas.

Also:

Ben Hargreaves at Lancashire Wildlife Trust.

Natural Resources Wales (downloaded March 2022). Stackpole National Nature Reserve Species Inventory and ad hoc sightings from across Pembrokeshire. Occurrence dataset accessed through the NBN Atlas.

Gloucestershire Centre for Environmental Records (downloaded March 2022); Gloucestershire Historic Wildlife Sightings prior to 1 January 2000. Occurrence dataset accessed through the NBN Atlas.

Buglife (downloaded March 2022). Invertebrate records from sites that are mainly in Scotland. Occurrence record dataset on the NBN Atlas.

West Wales Biodiversity Information Centre (downloaded March 2022). Natural Resources Wales regional data: all taxa (excluding sensitive species), West Wales; occurrence dataset accessed through the NBN Atlas.

South East Wales Biodiversity Records Centre. Natural Resources Wales regional data: southeast Wales non-sensitive species, SEWBReC (2018).

Woodmeadow Trust (downloaded March 2022). Woodmeadow Invertebrate Survey 2014, 2015, 2016. Occurrence dataset on the NBN Atlas; **Dave Raffaelli.**

Scottish Wildlife Trust casual records for Scottish Wildlife Trust reserves – verified data (2020).

Downloads of occurrence data for maps:

Colletes hederae download: NBN Atlas occurrence download at https://nbnatlas.org. Accessed 8 February 2022.

Anthophora plumipes download: NBN Atlas occurrence download at https://nbnatlas.org. Accessed 10 Mar 2022.

Anthidium manicatum download: NBN Atlas occurrence download at https://nbnatlas.org. Accessed 9 February 2022.

Heriades truncorum download: NBN Atlas occurrence download at https://nbnatlas.org. Accessed 10 Mar 2022.

Xylocopa violacea download: NBN Atlas occurrence download at https://nbnatlas.org. Accessed 10 10 March 2022.

IRISH DATA: National Biodiversity Data Centre, Ireland

Anthidium manicatum: accessed 22 February 2022: https://maps.biodiversityireland.ie/Species/TerrestrialDistributionMapPrintSize/55886.

Colletes hederae: accessed 22 February 2022: https://maps.biodiversityireland.ie/Species/TerrestrialDistributionMapPrintSize/56215.

Xylocopa violacea: accessed 22 February 2022: https://maps.biodiversityireland.ie/Species/TerrestrialDistributionMapPrintSize/57285.

P.125: *Halictus rubicundus*: Miocene warming: **M. Steinthorsdottir et al.** (2020), 'The Miocene: The Future of the Past', *Palaeoceanography and Palaeoclimatology*, 36 (4), https://doi.org/10.1029/2020PA004037.

Pp.125–126: Temperature and plasticity: **Schürch, R., Accleton, C., Field, F.** (2016), 'Consequences of a warming climate for social organisation in sweat bees', *Behavioral Ecology and Sociobiology*, 70:1131–1139 DOI 10.1007/s00265-016-2118-y.

Rebecca A. Boulton, Jeremy Field, (2022), 'Sensory plasticity in a socially plastic bee', *Journal of Evolutionary Biology*, Volume 35, Issue 9, 1 September 2022, 1218–1228, https://doi.org/10.1111/jeb.14065

Pp.137–138: For a taxonomic visual guide see the **Integrated Taxonomic Information System** website: https://www.itis.gov. The **Tree of Life Web Project** site is also worth exploring: http://tolweb.org/Apidae/22101.

Pp.141–142: Cardinal S. and Danforth B. N. (2013), 'Bees diversified in the age of eudicots', *Proceedings of the Royal Society B*, 280: 20122686: http://dx.doi.org/10.1098/rspb.2012.2686.

Also see: **Hedtke, S.M., Patiny, S., Danforth, B.N.** (2013), 'The bee tree of life: a supermatrix approach to apoid phylogeny and biogeography', *BMC Evolutionary Biology*, 13: 138.

Michez, D., De Meulemeester, T., Rasmont, P., Nel, A., Patiny, S. (2009), 'New fossil evidence of the early diversification of bees: *Paleohabropoda oudardi* from the French Paleocene (Hymenoptera, Apidae, Anthophorinae)', *Zoologica Scripta*, 38(2), 171–181, March 2009.

Ohl, M., Engel, M.S. (2007), 'Fossil history of the bees and their relatives' (in German: 'Die Fossilgeschichte der Bienen und ihrer nächsten Verwandten (Hymenoptera: Apoidea)'), *Denisia*, 20 (66): 687–700.

Pp.151–180: Mid legs of males of *Anthophora* species from around the world have been redrawn from illustrations in **Brooks, R.W.** (1988), 'Systematics and Phylogeny of the Anthophorine Bees (Hymenoptera: Anthophoridae; Anthophorinae)', *University of Kansas Science Bulletin*, 53 (9): 436–575, and **Brooks, R.W.** (1999), 'Bees of the genus *Anthophora* Latreille 1803 (Hymenoptera Apidae Anthophorini) of the West Indies', *Tropical Zoology*, 12 (1): 105–124.

ENDPLATES: a 'face-off' between males of *A. plumipes* and *A. hispanica*. You won't find the latter in a UK garden, though! This drawing was made from a photograph by Ian Cross on Flickr: https://www.flickr.com/photos/153642180@N07/albums/page4. I couldn't resist making a drawing of him and his extremely hairy legs and setting it next to our own hairy-footed male.

AFTERWORD

It is not yet spring when *Anthophora* flies, but for me, as a city dweller, her speed, her insistence on action, the energy bursting from her rather small form is a little like the start of lambing – a reminder that life renews itself and the excitement that comes with that. Or it is like the arrival later in spring of swifts over her city gardens. As the birds burst across the skies she is still working at the last of her nests.

Anthophora seems to be doing quite well and might be the easiest solitary bee to see in your local city park. There will be quite a few of her and she is, like others such as *Anthidium*, spreading northwards. The swifts, though, are not doing so well. Our skies are almost silent of them in places where before their abandonment to their wings, their turning of the sky into pure exhilaration has been an emotional anchor. Now they are struggling against our appropriation of Earth's space, the corruption of the skies, the destruction of their food and, by our increasingly powerful and pervasive wireless networks, the disruption of their magnetism. *Anthophora*, for the moment at least, seems to be cannily working herself into those spaces we are content to maintain – our gardens.

ACKNOWLEDGEMENTS

It's taken a while for this book to arrive in printed form: I began gathering ideas and drawings together in 2010. In the spring of 1999 I had taken on the management of a community wildlife garden for the charity Roots and Shoots, which is devoted to vocational training in horticulture and life skills for disadvantaged teenagers in Lambeth, south London. The encouraging management style of the charity (by Linda Phillips, director, and the trustees) gave me a lot of leeway to design not just garden management principles for the 1.5 acres of what would otherwise be a prime central London development opportunity, but also educational programmes to spread the word about the delights to be found in the urban natural world. Early on I was greatly helped by the practical and moral support of the late Clive Watson, a remarkable beekeeper and shower fitter. As swarm collector for south London, he could be spotted in his van driving down the Walworth Road with stray worker bees in dogged pursuit! Clive, and later, artist Charlie Millar, worked with me on the apiary in the Wildlife Garden, and with the London Beekeepers Association, the Central Association of Beekeepers and Bob and Mary we ran beekeeping courses at Roots and Shoots from 2000–2010. I was convinced from the beginning, however, that studying and communicating the diversity of the bees in the wildlife garden was crucial. Over 21 years we ran hundreds of meetings, courses, lectures and events, from apple days (thank you, David and Gayle) to bee walks, and hosted a large diversity of 'bee folk', both academic and practical. The work needed to organise these events –the observations, recordings, documentation and research – generated resources that could bring awareness of biodiversity to all our audiences, including beekeepers. The photographs and drawings that I used were the beginnings of this book. Significant funding in that time came from City Bridge Trust, the Walcot Foundation and The Sheepdrove Trust.

The book has had more than one format, but from the beginning I have been supported by Norman Carreck, who first visited Roots and Shoots when he was still at Rothamsted Research. He was very interested in my first, family-learning version of the book whilst senior editor of the *Journal of Apicultural Research* and science director for the International Bee Research Association. After a lot more drawing and advice from Norman and others it has reached the form you now see. I am very grateful to Keith Whittles, who took the book up at a time (just after the pandemic) when I thought it wouldn't make it, and to his team who have helped to bring it to its final state. It might appear a difficult book to categorise, but from whichever angle you are approaching, I hope you find its current form stimulating and enjoyable.

I would like to thank the staff of the many biological and environmental records centres for allowing me to use the relevant data for the section on bees and climate change, particularly Úna Fitzpatrick at the National Biodiversity Data Centre, Ireland and Ben Hargreaves of the Lancashire Wildlife Trust. Professor Dave Raffaelli, then chair of trustees at Woodmeadow Trust was very welcoming and hospitable at the Trust's site in Yorkshire. Sarah Wilson, gardener and artist at Roots and Shoots, has been a loyal friend and supporter.

I wouldn't be where I am without Mary Jo.

Anthophora plumipes

Anthophora hispanica

NOTES:

Hairy-foot long-tongue

Of all the bees that you might see in your garden, how many of them are NOT honeybees or bumblebees? More than you might think.

This book introduces the solitary bees: bees with beautifully adapted bodies and life cycles as interesting as the more familiar honeybees.

Guided by the lovely early spring bee, the HAIRY-FOOTED FLOWER BEE, we find out why bees are hairy, why they have long tongues, how and where they build their nests and some of the fascinating details of their adaptations to living in the habitats of our gardens.

With hundreds of the author's drawings and photographs we explore the natural science of these bees in an accessible and attractive format.